Geosonics

Geosonics

Listening Through Earth's Soundscapes

Joshua Dittrich

BLOOMSBURY ACADEMIC
NEW YORK • LONDON • OXFORD • NEW DELHI • SYDNEY

BLOOMSBURY ACADEMIC
Bloomsbury Publishing Inc, 1359 Broadway, 12th Floor, New York, NY 10018, USA
Bloomsbury Publishing Plc, 50 Bedford Square, London, WC1B 3DP, UK
Bloomsbury Publishing Ireland, 29 Earlsfort Terrace, Dublin 2, D02 AY28, Ireland

BLOOMSBURY, BLOOMSBURY ACADEMIC and the Diana logo are trademarks of Bloomsbury Publishing Plc

First published in the United States of America 2024
Paperback edition published 2026

Cover design: Louise Dugdale
Cover image © NASA "blue marble"/ Visible Earth

Christine Sun Kim, *The Sound of Laziness*, 2016. Charcoal on paper, 19.5 x 25.5 inches (49.5 x 65 cm). Courtesy of the Artist and François Ghebaly Gallery, Los Angeles. Photo: Kell Yang Sammataro.

Christine Sun Kim, *The Sound of Apathy*, 2016. Charcoal on paper, 19.5 x 25.5 inches (49.5 x 65 cm). Courtesy of the Artist and François Ghebaly Gallery, Los Angeles. Photo: Kell Yang Sammataro.

Christine Sun Kim, *The Sound of Being Resigned*, 2016. Charcoal on paper, 19.5 x 25.5 inches (49.5 x 65 cm). Courtesy of the Artist and François Ghebaly Gallery, Los Angeles. Photo: Kell Yang Sammataro.

Christine Sun Kim, *Degrees of Deaf Rage within Educational Settings*, 2018. Charcoal and oil pastel on paper, 49.25 x 49.25 inches (125 x 125 cm). Courtesy of the Artist, François Ghebaly Gallery, Los Angeles, CA, and White Space, Beijing. Photo: Yang Hao.

Christine Sun Kim, *The Sound of Temperature Rising*, 2019. Acrylic on wall adapted from 2016 charcoal on paper drawing, Edition of 3, 1AP, 115.75 x 196.75 inches (294 x 499.74 cm). Installation view, PS120, 2019, Berlin, Germany. Courtesy of the Artist and François Ghebaly Gallery, Los Angeles. Photo: Stefan Korte.

Christine Sun Kim, TBD, TBC, TBA, 2015. Charcoal on paper, 11 x 15 inches (28 x 38 cm). Courtesy the Artist, White Space, Beijing, and François Ghebaly, Los Angeles. Photo: Yang Hao.

Sun Ra's poem "The Empty Space" appears courtesy of Sun Ra LLC and Corbett vs. Dempsey.

Library of Congress Cataloging-in-Publication Data
Names: Dittrich, Joshua, author.
Title: Geosonics : listening through earth's soundscapes / Joshua Dittrich.
Description: [1.] | New York : Bloomsbury Academic, 2024. | Includes bibliographical references and index.
Identifiers: LCCN 2024006306 (print) | LCCN 2024006307 (ebook) | ISBN 9798765104576 (hardback) | ISBN 9798765104569 (paperback) | ISBN 9798765104583 (ebook) | ISBN 9798765104590 (pdf)
Subjects: LCSH: Soundscapes (Music–Philosophy and aesthetics. | Music–Environmental aspects. | Sound–Environmental aspects. | Sound (Philosophy)
Classification: LCC ML3877 .D43 2024 (print) | LCC ML3877 (ebook) | DDC 80.1—dc23/eng/20240322
LC record available at https://lccn.loc.gov/2024006306
LC ebook record available at https://lccn.loc.gov/2024006307

ISBN: HB: 979-8-7651-0457-6
PB: 979-8-7651-0456-9
ePDF: 979-8-7651-0459-0
eBook: 979-8-7651-0458-3

Typeset by RefineCatch Limited, Bungay, Suffolk

For product safety related questions contact productsafety@bloomsbury.com

For Daniel and Martin

&

For Stanka

Contents

Illustrations

Introduction

Is there an audio equivalent of the famous *Blue Marble* photograph of earth, taken from the Apollo 17 spacecraft in 1972 (Figure 0.1)? This image encapsulated the technological ambition of the space age as well as the geopolitics of the Cold War, and prompted a paradigm shift in ecological consciousness that inspired the environmental movements of the 1970s. Is it possible to imagine sound and listening impressing the planet "itself" on the human imagination with the same intensity as an image like this?

One technical, if somewhat *boring* answer is: yes. Recordings of the vibrations or "sounds" of earthquakes have long played a crucial role in understanding and representing the deep structure of the earth. Starting at least as early as 1909, when Croatian seismologist Andrija Mohorovičić defined the gap between the earth's crust and mantle by listening to patterns in earthquake vibrations (a technique of listening I discuss in Chapter 2), seismologists have understood the earth by picturing it in sound. Likewise, an unexpected benefit of the military use of sonar technologies in the submarine warfare of the Second World War was the capacity to map the seafloor, generating unheard-of data about underwater mountain ranges and trenches that would, by the late 1960s, help to make the case for plate tectonic theory.[1] So, yes, it's fair to say that earth scientists have used sound and technical listening to visualize the earth, and in fact, had made a regular practice of doing so for decades before the spectacular *Blue Marble* photo was taken.

Another possible answer (also *boring*, but in a different way, as I suggest below) is: no, not exactly. Listening to sound is a dynamic, relational process: it implies the sharing of space and time across the sound's source, the listener, and the sonic environment. Listening demands the sustained and highly subjective attentional work of perceiving and interpreting sound, constantly (re)locating it in relation to the self, the source, and the environment, and taking into account how each of these listening elements might be modified or transformed altogether in the process. Listening also raises questions of scale and mediation:

Figure 0.1 *The Blue Marble* (taken by the Apollo 17 crew in 1972) NASA.

what are the material thresholds by which a sound is heard in the first place? And what are the cultural or aesthetic thresholds that make a sound meaningful for a listener at a given place and time? If so much of listening necessarily involves *not listening* to all the sounds that don't pass the material and cultural thresholds of our attention, how do technical means redefine our attention to listen to the sounds we can't otherwise hear?

The *Blue Marble* certainly raises questions about technical mediation, too—What kind of camera and film stock did the astronaut-photographers use? How do you adjust the aperture and exposure when your light source is the sun itself, unfiltered by earth's atmosphere? How big is a lens that will hold an object in focus at a distance of 29,000 kilometers?—but we suspend such questions

according to the self-evident logic of the image, which invites us to imagine that an astronaut simply pointed a camera out the back window of a spaceship, and this is what he saw. Sound, it would seem, just can't exude "planetariness" in the same iconic, immediate way. So how does sound help us to listen to the planet? What does the earth sound like?

Before I take on these questions in the chapters that follow, let me offer a brief story here as a preliminary answer. A few months ago, I was driving to the suburban campus where I teach, mulling over how to get started on this book's final chapter on composer and bandleader Sun Ra's "astrosonic" musical practice. In the car I was listening to a recording of Ra and his band, the Arkestra, play a piece called "The Voice of Space" (I discuss this track in Chapter 6), but my mind was wandering back to a scene from Ra's 1974 film *Space is the Place*, which I had just re-watched the day before. In the scene, Ra (playing a version of himself) works a counter at the "Outer Space Employment Agency" in Oakland, California, to try to recruit personnel to help him transport Black people to a distant planet. As the camera moves from the street into the Agency at the start of the scene, we hear the Arkestra on the soundtrack playing a catchy, upbeat tune, with singer June Tyson crooning, "If you find earth boring / just the same old, same thing / Come on and sign up / with Outer Space Ways, Incorporated." It is a hilarious and often discussed set piece within the film, comically translating space age rhetoric into the languages of radio advertising and sketch comedy gags.[2] But it wasn't helping me to get a grasp on "The Voice of Space" (which, by contrast, is a very difficult improvised piece—and terrible music for driving, by the way!), and I was struggling to reconcile the forbidding, idiosyncratic virtuosity of what I was hearing on the car stereo with the playful sensibility of the film music that was replaying itself in my head like an earworm.

And suddenly, right then and there, I found earth boring. To be more exact, I saw a sign that said "Earth Boring" on a chain-link fence around a construction site next to the highway (Figure 0.2). Earth Boring, it turns out, is the name of a construction company based in the same suburb where I teach, specializing in underground projects like tunneling, pipe ramming, and something called auger drilling. Of course, I thought as I drove on, having snapped my own picture from the window of my spaceship/car: *boring*, via its Germanic roots, means drilling, piercing, or perforating. And earth boring, I then recalled, is a technique of geological sampling otherwise known as sounding, from the Germanic roots of "sound," which likewise means to penetrate or pierce. Depth soundings in the water refer to how far a plumbline will "sound" or sink into the water until it hits

Figure 0.2 Earth Boring. Author photograph.

the bottom. Earth scientists use soundings to measure not only rock and water, but air as well: weather balloons sound the atmosphere, boring up into the air, as it were, to gather data about the pressure, temperature, and composition of the planet's outer reaches.[3]

I had found earth boring, indeed, insofar as I encountered the same fragment of language twice, but in two totally different contexts. The same words not only took on different meanings, but they also opened up a conceptual and etymological "rhyme" (through shared notions of measuring by piercing) with sound. From there I began to listen to the music still playing on the stereo differently: in Ra's "piercing" high notes on the Hammond organ, I imagined a sounding device boring into the sonic space created by the rest of the band, and that the reverberant echo effects used on the recording were a kind of echolocation, with the titular "Voice of Space" exploring, measuring, traveling into sound and by means of sound. These kinds of metaphors are, of course, central to Ra's rhetoric of "space music," but I could now grasp how literally they were grounded in geological techniques *and* in the etymology of the words that describe those techniques. Meanwhile, they also corresponded in an oblique, but compelling way to the two soundscapes that composed my daily commute: the difficult, improvised music on the stereo, and the earworm "If you find earth boring" that I couldn't get out of my head. "Earth boring" was a hinge that opened on to an unexpected continuity between actual and imagined sounds; technical and imaginative modes of listening; the ambiguity of words; and the earth itself, that boring site right there on the other side of the fence.

Listening to the Earth

Geosonics explores the materialities of sound and the technical and imaginative dimensions of listening on a planetary scale. As I suggest in the above anecdote, "planetary" does not necessarily imply vastness or immensity. We can listen to the earth, or imagine that we do so, in subtle, unexpected ways. It's a small world after all, and *Geosonics* attends to the circuitous, mediated ways in which the planetary, the technical, and the aesthetic merge with, but also interrupt the soundscapes of everyday listening, language, and imagination.

The question, "How do we listen to the earth?" is a sound studies question, of course, because it concerns listening, which I approach via the techniques and tools that determine what counts as sound, and which co-evolve with embodied and encultured practices of space-making in and by sound. It is also a media-theoretical question because it considers mediation as a fundamental process that underlies and gives rise to subjects (like the listening "we") and objects (like the "earth"). And it is an environmental media question because the climate of contemporary intellectual work cannot help but register the effects of the literal climate of the planet. Global warming, resource depletion, species extinction, and a generalized sense of planetary finitude (exacerbated by the Covid pandemic and the renewal of nuclear end-game scenarios caused by the Russian invasion of Ukraine in 2022) are transforming how scholars think about theory, culture, and technology across disciplines. In *Geosonics*, climate imposes an inevitable, if multi-scalar materialism that yokes together humans, technology, and geology itself.

"We," as subjects, are effects of, or entities distributed within, mediation as a kind of ontological environment: media-wise, we are always "on." We are also always *on the earth*, which is the ultimate environment, or the environment of environments. *Geosonics* explores precisely this congruence of mediation and earthliness: how mediation is geologic, how geology is mediated, and how the instruments and bodies we use to sense our way, aesthetically and technically, into mediation are, every step of the way, interfaces with the earth. From a humanities and media-studies perspective, the more we attempt to think about the planet "itself," the more we are actually grappling with a complex problem of mediation: the conversion of earth-scale phenomena into objects that humans can sense, and make sense of.

Within this aesthetics and geology of mediation, *Geosonics* argues for the critical and epistemological centrality of sound and listening. In Chapter 1, I characterize sound as a vibratory and temporalizing dimension of matter. Sound is the process of matter in mediation, and in that sense, has a privileged status in the geologically inflected materialism I am interested in. This is not to say that I privilege listening explicitly against—or in isolation from—other senses; indeed, I tend to approach all the senses always as an embodied ensemble. But I do insist on sound as a key pathway for theorizing sensation and embodiment more broadly in a robustly reciprocal set of relations between bodies, media, and the sites or environments in which sensation occurs. In an epoch of climate crisis, however, environment is no longer a neutral background, site or a simple "surrounding": environment itself is also another constitutive link in the chains of mediation that make up the material and imaginative infrastructure of our lives. The analytical task of *Geosonics* is to tune into that infrastructure through sound. Drawing on influential work in sound studies around the concept of transduction (a concept that refers to the audio conversion of sound waves and electrical signals, but can also more broadly describe the mediated conversions between material and cultural processes), I develop a practice of transductive listening: to listen *through* the earth's soundscape, literally and imaginatively boring through the immersive self-evidence of the sonic experience of the environment (qua soundscape, nature, background, atmosphere, ambience, world, etc.) to attend to the earth as a technical and cultural artefact whose materiality resounds in the imagination.

The difficulty with transduction is that, when it works, the infrastructure underlying it disappears experientially. Transduction makes the environment-as-mediation disappear into the immediacy of environment-as-sensation. In order to render explicit the infrastructural work of mediation and geology in transduction, I take a twofold approach throughout the book: at times, deconstructing experiences of sonic immersion into their constitutive transductive parts (body, media, earthly interface), at other times revealing hidden continuities between seemingly "small" acts of listening and the vibratory dynamics of the planet. *Geosonics* attends to the material infrastructures that allow us to imagine our immersion in a sonic environment, and to the ways in which embodiment—the complex, processual experience of being-in-a-body—is inseparable from the experience of being-on-a-planet. One central aim of *Geosonics* is to use sound and listening to expand how scholars think of materiality at human and geologic scales; and to argue for the centrality of transduction as a critical interdisciplinary tool for rethinking the complex relationships between earth, media, and experience in the twenty-first century.

Listening through Soundscapes

Since the notion of soundscape is consistently (if at times only implicitly) bound up with my approach to sound and listening in *Geosonics,* I first need to contextualize my project against that concept's evolution. The soundscape is synonymous with R. Murray Schafer's project of acoustic ecology, as set forth most prominently in the 1977 book, *The Soundscape: Our Sonic Environment and the Tuning of the World.* A soundscape is an acoustic environment *and* a field of study simultaneously. It involves the study of the unique sonic profile of a given environment as a means of disclosing a certain knowledge of that place accessible only in acoustic terms.[4] It is the sonic equivalent of a landscape (as I discuss further in Chapter 1), but the term can also apply to acoustic design in built environments, to musical performances, works of sound art, and even radio broadcasts. What is most decisive for Schafer is that the soundscape is both a broad aesthetic and analytical concept (opening onto an array of musical, sociological, and architectural/design modes of engaging and creating in/with the sonic environment) *and* an embattled, evaluative term at the same time. Soundscapes usually imply, for Schafer, their own degradation. As modern life grows increasingly populated with mechanized and commercially recorded sounds, the ability to listen declines in proportion, and that aesthetic loss, for Schafer, entails a kind of moral and spiritual loss as well. Schafer combines an ancient Pythagorean version of "the music of the spheres" (the mystical contemplation of the numerical-tonal unity of the cosmos) with a Romantic, anti-capitalist idealization of nature and contemplative silence. It is a paradox that the soundscape project employs the tools and technical language of modern auditory culture to map an increasingly heterogeneous soundscape, while at the same time cleaving to a relatively homogenous, normative ideal of natural, environmental sound and "clairaudient" listening practices.[5] Furthermore, the soundscape concept as Schafer developed it, retains explicit settler-colonialist and Canadian nationalist overtones that problematically tokenize, marginalize, or exclude altogether the musical and sonic life of non-Europeans from the picture.[6]

Despite—and perhaps because of—its limitations and contradictions, the soundscape concept has been generative for critical work in the field of sound studies, in addition to continued work in the acoustic ecology paradigm.[7] Brandon LaBelle describes his approach in *Acoustic Territories* as an attempt to "supply the notion of soundscape with an overt form of tension and political charge" ([2010] 2019: vi), working explicitly against the anti-urban bias of the soundscape concept to explore how sound contours the spatial politics of contemporary urban life. Allie Martin's ongoing work across the fields of digital humanities and ethnomusicology uses soundscape recordings and digital tools, which recall earlier acoustic-ecological practices of sound mapping, to analyze sound as a vector of gentrification in Black neighborhoods of Washington, DC. Her recordings and visualizations of the sounds of traffic, construction, sirens, and overheard go-go music (all of which might be dismissed, in varying degrees, as noise in Schafer's terms) create an archive of the vitality of Black sonic life in DC, politicizing the soundscape in anti-racist terms.[8] From an ethnographic angle, literary scholar Edwin C. Hill Jr. has reinvented the concept of soundscape as soundsc(r)ape to critically examine sound's inscription in racial hierarchies that structure ethnographic and literary writing about and in the Afro-Caribbean diaspora. For Hill, "[t]he matter of soundscapes must [. . .] include the sonorous black bodies and cultures written into, and out of, European systems of meaning and value" (2013: 14). Updating the soundscape concept means taking the tools and methods of acoustic ecology (which, it must be noted, played a crucial role in the emergence of sound studies and in legitimizing sound and listening as interdisciplinary scholarly objects), but working against the preservationist, normative worldview that permeates them, and adapting them as critical epistemological tools to describe sonic diversity in open-ended ways.

In one of his most cited remarks from *The Tuning of the World*, Schafer describes "the soundscape of the world as a huge musical composition, unfolding around us ceaselessly. We are simultaneously its audience, its performers, and its composers" (1977: 205). In a similar vein, looking back on her influential work with Schafer in acoustic ecology, Hildegard Westerkamp cites acoustician Michael Stocker to affirm that "'our experience of sound unfolds as a continuous now'" (2019: 45). The rhetoric of immersion, continuity, presence, and musicality is crucial to acoustic ecology's notions of soundscape, along with that project's cultivation of inner, awareness-raising practices of listening that tune listeners into their environments in aesthetic and (more or less explicitly) spiritual terms. *Geosonics*, by contrast, attends to what is roughly the same sonic field, the earth-scale "soundscape of the world," but in discontinuous, fragmented, repetitious, and explicitly mediated terms. Listening to the earth is a transductive process, which means it involves the constant conversion of energies, matters, and meanings across unstable thresholds that say more about the technical aspects of bodies, cultures, languages, and media than they reveal about the earth "itself." To anticipate an argument I make in Chapter 1, soundscapes are sound*ing*scapes, which disclose—and afford the opportunity to critically attend to—the technical work of mediation and imagination that composes the sonic environment.

Eco or Geo? A Note on Sonic Essentialism

Having situated *Geosonics* on the other side of the soundscape, I now need to address the conspicuous absence throughout the book of environmental or ecological sound art. These terms refer to works of sound art and music composition which use field recordings, geophysical data, and sonification to create an aesthetic connection between listener and environment, and to engage environmentalist issues such as climate change, biodiversity loss, and environmental justice. Such artworks involve complex technical negotiations of media and scale to establish their connections, and depend upon the premise that sound possesses inherent capacities to communicate ecological meanings and to cultivate individual and collective agency regarding environmental issues.[9] One often cited example is Scottish artist Katie Paterson's 2007–8 installation *Vatnajökull (the sound of)*, in which the artist constructed a live phone connection to a microphone/hydrophone underneath a melting Icelandic glacier. Listeners around the world could call a phone number and listen to the sound of the glacier melting in real time. Speaking of this project in a 2021 TED talk about her work titled "The Mind-Bending Art of Deep Time," Paterson explains that in *Vatnajökull*, she "wanted to bring these distant landscapes closer to our lives in a visceral way," and that the piece is "an elegy to a disappearing landscape." At this point in the talk we hear a few seconds of watery-sounding audio, presumably recorded from the original installation.

From one perspective, the technical set-up of this exemplary work of environmental sound art is astounding: direct audio access across vast tracts of space and across media (meltwater, hydrophone, telephone) allows listeners to hear for themselves the sound of global warming in real time and via everyday media devices, providing the "visceral" reminder that the remote effects of climate change are closer to home than we think. But from another perspective, the installation reproduces a number of cultural assumptions about nature, media, and listening. The piece assumes that "nature" is out there, immediately available to human ears, and that microphones and phone lines are neutral media that unproblematically amplify our inherent connection to the natural world. It also assumes that nature will sound like an elegy, i.e., that the audio material's naturalness and poignancy will be guaranteed by the implied listening framework of a Western poetic and musical genre. The elegiac sounds are also "visceral," and Paterson must have made a number of aesthetic and technical decisions about how to process the underwater sounds to make them sound as compelling and climate-changey as possible. In short, aesthetic preconceptions about sound and listening abound here in this phone call to climate change, while the rhetoric of environmental sound art filters out the cultural and technical transductions that in fact constitutively frame the installation's premise of immediacy.

In *Geosonics*, I want to critique the rhetoric of immediacy in ecological sound art because it prevents listeners from attending to—and questioning—the specific epistemological and technical work that "natural" sounds and soundscapes do when they are presented to us as experiences of immersion and of "nature." In this regard, the critical aspect of *Geosonics* finds support in Frances Dyson's argument about immersion in *Sounding New Media* (2009). Dyson argues that new media and digital technologies dial in a rhetoric of immersion and interactivity borrowed from an uncritical metaphysics of sound (as immersive, primordial, omnipresent, transcendent, et cetera), in order to present (indeed, to market) online and virtual environments as immersive realities to users/consumers. Sound, as Dyson points out, has a complex philosophical history of uncontainable, problematic relations to presence, the body, and communication; but that has not stopped contemporary technocultures from instrumentalizing sound as the hidden supplement to immersive, imperceptibly mediated experiences of presence, starting from the "old" media of telephone, radio, and cinema, and extending to the new media of virtual reality. The crucial point for Dyson is that "although sound as a material acoustic medium provides an opening to an alternate metaphysics, it does not in itself constitute that metaphysics, and it would be a mistake to conflate sound, or even the larger sphere of aurality, with the alterity it points to" (5).

Mapping Dyson's argument onto the aesthetics of ecological sound art in Paterson's piece, we can see the conflation of a seemingly self-sufficient, "natural" sound (e.g., of a glacier melting) with the radical, unheard-of alterity of climate crisis and the radical forms of agency that it necessitates. Rather than asking us to imagine how we could possibly listen to what climate change sounds like (an extraordinary, arresting question), the piece tells us "This is what climate change sounds like," and the answer it provides is familiar and utterly continuous with our everyday media experience. The gesture to make climate change "visceral" through sound also makes climate change *virtual*, an instantly recognizable figment of circuits and digital signals produced by a society so committed to technologies of instantaneous communication that it destroys ecosystems and melts glaciers to maintain 24/7 connectivity. The phone call to the glacier is, despite its intentions, structured like any telephonic communication and any online service, and risks reinforcing the infrastructural banality of those (ecologically wasteful and destructive) communications networks along with the environmentalist message it rightly wants to convey. In contrast to ecological sound art's premise that sound is "an inherently ecological medium—one that enables a direct sensorial engagement with the agency of vibrant matter, facilitating an embodied understanding of our embeddedness within an interconnected ecosystem" (Gilmurray 2017: 40), I engage with sound as repetitive, interruptive, and indeterminate. And I conceive of listening as a discontinuous, looping process of remediation across material, cultural, semiotic, and imaginative registers that replays sound's reverberant nature as a modality of critical and self-reflexive attention. Though I fully support the environmentalist pedagogy and politics of environmental sound art (and in fact frequently teach pieces by Katie Paterson, Leah Barclay, Matthew Burtner, and others—not to mention key texts and soundwalks associated with acoustic ecology), I want to explore in *Geosonics* the works of sound art, pieces of music, and the listening practices which hold open space for indeterminacy and alterity, and which resist the smooth rescaling of the geological to suit the human picture of nature.

On the concept of scale, Zachary Horton's new materialist approach in *The Cosmic Zoom* (2021) is illuminating: humans, for Horton, occupy a "multi-scalar flux" (18) and are themselves "scale-unstable" (7) entities, entangled in vast structures, systems, and rhythms (like climate change, Big Data, and capitalism) whose complexity defies comprehension. Humans, at the same time, constantly build media infrastructures and disciplinary knowledge practices to homogenize and naturalize that which is radically other to their sensory and epistemological scales. What is aesthetically crucial for Horton is the "trans-scalar encounter," the gap or lull between an experience of scalar alterity (that point of contact with an unknown and, for the moment, inassimilable modality of experience or knowledge) and the reassimilation and rescaling of alterity as essentialist, techno-determinist business as usual. In *Geosonics,* I translate such trans-scalar encounters (which Horton analyzes primarily in literary, cinematic, and digital media contexts) into a critical-materialist study of soundscapes. How do media-technical artworks suspend their in-built logic of assimilation and hold open space for a different kind of thinking and knowing? What does climate change sound like in the moment *before* it becomes a phone call to a glacier? And how do we continue to listen to that sound once its un-heard-of-ness has been absorbed into the everyday banality of global telecommunications?

Melody Jue's *Wild Blue Media* (2020) provides a valuable counterpart to the immersive, essentialist thinking about nature and sound that I attribute above (and with an admittedly heavy hand) to ecological sound art. Jue unsettles media essentialisms of the environment by taking media underwater. Her notion of "milieu-specific analysis" approaches the concept of environment not by connecting to nature, but by defamiliarizing it, attending to the conditions of perception and techniques of mediation through which a natural environment is constituted in the first place. In shifting her analysis from terrestrial to marine environments and thinking *through* seawater (that is, through real or imagined bodies of sea creatures, or through the embodied experience of technical interfaces [like scuba gear] that allow humans to negotiate a watery environment), she enacts a conceptual displacement that draws our "terrestrial bias" into relief: how much of human thinking presupposes sitting at a desk, standing, walking, as grounded modes of engaging media and the environment, and what happens when we can't take gravity and ground for granted? We come to know differently the (terrestrial) media world that we already inhabit by thinking through media and mediation underwater. Citing Donna Harraway, Jue locates her environmental media perspective in "the always partial, always finite, always fraught play of foreground and background, text and context, that constitutes critical inquiry. Above all, location is not self-evident or transparent" (Harraway, cited in Jue 2020: 37). For Jue, location—whether it is a critical, epistemological position, a place on the earth, or an underwater milieu—emerges as a mediating process of sounding and listening.

Geosonics enacts a similar conceptual displacement to Jue's work by thinking the earth through sound, activating critical and aesthetic awareness of the earth as an ongoing process of (sonic) mediation. From this perspective, the soundscapes of acoustic ecology and ecological sound art share a "terrestrial bias." Geosonics, by contrast, seeks to deconstruct the "earth" that is the phenomenological and ideological horizon of the soundscape, merging the background, foreground and underground conceptually and sonically, to render the entire soundscape strange, partial, mediated, and full of unexpected temporal, spatial, material, and linguistic juxtapositions.

Sounding the Structure of *Geosonics*

In listening through soundscapes, *Geosonics* rethinks the relations between listening and environment along three recent paradigm shifts in the humanities: (1) new media and digital studies, i.e., attempts by humanities scholars to critically understand all the ways that computation and digital media have come to reshape knowledge and society; (2) new materialism, post-human, and speculative perspectives that perceive new kinds of relationships and potentials distributed across networks or assemblages of humans, objects, and ecologies; (3) environmental media studies which, by measuring the material footprints of media production and consumption, critically questions how humans continue to make knowledge in an epoch of anthropogenic global warming. Each chapter develops arguments around specific technical listening practices, artworks, and music performances, while engaging particular texts and problematics in digital, new materialist, and environmental-media approaches to sound.

What ties the chapters together is a certain split or delayed temporality, which, I argue, is built into the structure of listening. As intimated in the "earth boring" story, the premise of *Geosonics* is that we hear every sound at least twice, and that the soundscape of the world is not a "continuous now," but a complexly mediated reverberation that changes the sonic landscape precisely when we try to compose it for ourselves. I offer a more detailed discussion of the "always again" structure of listening in Chapter 1, but it starts off here with some simple facts of how humans usually perceive sound. Audition in space is binaural: as a sound moves through space, it will activate one ear, then—after an infinitesimal lag as the sound travels through the denser matter of the skull—the other ear. This delayed or doubled hearing sets up a kind of neuro-triangulation that allows the brain to locate sounds in space, to confirm that the bicyclist behind you is indeed coming "on your left!" If you have two hearing ears, then each time you perceive a sound, you perceive it twice, and your own head both obstructs and finetunes the process.

Moreover, the field of Deaf studies has drawn into focus how listeners also experience sound through an ensemble of tactile, proprioceptive, and visually inferred modalities, contracting multiple acts and sites of sensation (in the bones, skin, muscles, and in perceived visual and social patterns) into the hearing-again of a "single" sound.[10] Likewise, when you hear a sound, you also hear the *reverberations* of that sound at nearly the same time. Some energy of a given sound travels "directly" through the air to your ears/body, but most of it will bounce off all the other available surfaces along the way and arrive obliquely at your perceptual doorstep a fraction of a second later. Sound thus constantly re-originates and repeats itself as it moves. Hearing "a" sound necessarily means hearing a reverberant composite of sounds that vexes the distinction between a source and its echoes. Put another way, reverberation means that whenever we hear a sound, we also hear the sound of the surrounding space (via the particular qualities of the local acoustics and even of the air itself) as it fills up with that "first" sound.[11] The act of perceiving a sound, then, is doubled (and then some) by sound's reverberation in and through space, even the inner space of the listener's skull.

Geosonics expands the scale of these details from physiology and acoustics to attend to how a certain doubling of sonic movement is built into more complexly mediated forms of listening. When we listen to sound as a physical stimulus, as material energy moving through space and affecting our bodies, we also register its impact on our aesthetically, technoculturally, and politically trained ears, as well as the idiosyncrasies of our individual, embodied hearing abilities. In registering the impact of the sound, we hear it *again* as a meaningful construct that elicits our attention or affective responsiveness in particular ways. The flipside of this same logic means that our trained ears co-determine what counts as attention-worthy sound in the first place. We hear a sound out there in the world because we have already heard it through the embodied apparatus of our particular auditory culture and lived experience. Chapter 1 plays out this feedback loop between the materiality and metaphoricity of sound in the context of new materialist media theory, exploring what I argue is a sonic-theoretical underpinning to much of recent thought on the matter of media. I develop a method of transductive listening by juxtaposing my take on Stefan Helmreich's noted essay, "An Anthropologist Underwater," with some examples of actual and imaginary drilling-into and listening to the earth, and against the microphonic musical aesthetics of John Cage and others.

Chapter 2 focuses on techniques of listening in seismology (the study of earth-scale vibratory movements, especially earthquakes), and how sound helps to compose a geologic picture of the earth. From the perspective of a seismometer, all earthquakes sound twice: the seismograph detects one distinct set of vibrations that travel as elastic waves through deep layers of rock, followed by a second, slower-moving set of vibrations that propagate as sheer waves along the surface of the earth. By analyzing the intensity of each set of waves as well as the time lag between them, seismologists can learn about not only the earthquake itself, but the composition of all the surrounding rocks, and eventually (as in the famous example of Mohorovičić mentioned above) the deeper structure of the earth. The doubled "sound" of earthquakes bears out an uncanny similarity to the doubledness of ordinary human listening, and this chapter explores the mutually constitutive epistemologies of sound and seismology in a range of technical and artistic earthquake sonification practices.

Chapter 3 takes on a paradox of contemporary neoliberal and consumerist auditory cultures, transforming a question like, "What if we had more time for listening to music?" (enter streaming services, noise-canceling headphones, AirPods, etc.), into, "What if we had more time for listening to music while we were already listening to that same music?" The central artwork discussed in Chapter 3 is a 24-hour version of Beethoven's Symphony No. 9 (titled *9 Beet Stretch* by Norwegian artist Leif Inge) that uses a digital technique known colloquially as "audio stretching" to extend the playback time of that famous piece without altering its pitch. As a result, listeners can devote roughly 22 seconds of their listening time to attend to each second of the original recording's running time, listening to the piece again and again while, at the same time, listening to it just once. The piece invites us to listen to the symphony as a 24/7 environment, playing with attentiveness, embodiment, and temporality on a planetary scale. Whether to read this experiment as resistance to the 24/7 consumerism of the contemporary digital world, or as its dubious crowning achievement in auditory terms, is the question I grapple with, thinking through the work of Jonathan Crary, Mark Hansen, and Susanne Cusick, alongside other durational artworks and performances.

Chapters 3 hinges on the premise that listening *through* the soundscape entails critically attending to the audio environments and auditory cultures produced not only by the earth, but also by the earth-scale, global system of extractivist capitalism and its dominant neoliberal ideology. In that sense, 24/7 connectivity to (audio) media composes a "natural," *ideological* soundscape of the early twenty-first century, so I argue that a 24-hour, cyclical audio installation like *9 Beet Stretch* can afford listeners a critical perspective because it shares the temporality and media infrastructure of the 24/7 world.

In Chapter 4, I examine the technology of the human senses as sites where the soundscape (both as "natural" environment and ideological surround) is converted from an external perception to an internal sensation. I focus first on the art and activism of cyborg artist Neil Harbisson, whose creative work derives from an antenna implanted in his skull that sonifies color frequencies. Totally color-blind since birth, Harbisson uses technology to hear a doubled or augmented soundscape of acoustic sounds and sonified colors, a prosthetic enhancement that he promotes as a self-appointed spokesperson for cyborg rights and an advocate for a trans-species cyborgism. In media studies, the cyborg has proven to be a figure for rethinking ability, prosthesis, and embodiment, but I argue that in Harbisson's case, cyborg becomes an identity position that aligns with neoliberal and transhumanist rhetorics of bodily enhancement. I use Adrian Mackenzie's work on transductive bodies to critique the seamless interfacing that Harbisson posits between environment, body, and technology. By critically attending to Harbisson's (mis)hearing of the sensory gap between color and sound, I am able to conceptualize the frictions and indeterminacies of the otherwise smooth twenty-first-century digital soundscape.

I conclude the chapter with a discussion of the work of Deaf artist Christine Sun Kim who, like Harbisson, works with sound as a means of centering what for her is an unavailable sensory modality. But, approaching that unavailability from a different angle, Kim uses sound to initiate a conversation precisely about sound's taken-for-grantedness and, in a way, *disappearance* into social and semiotic processes of remediation, and into the ideology of a hearing world. Transducing sound by its seeming absence, Kim's work establishes a listening framework in which deaf and hearing audiences can reflect on the culture of sound's collective affordances and failures, translating issues of communication, access, and ability into an aesthetic soundscape without sound.

In Chapter 5, I proceed from a split inherent in listening between conscious attention and nonconscious affectivity (i.e., how we can be affected by sounds even when we do not consciously listen to them) to explore the congruence between sleeping and listening as nonconscious processes. I approach sleep as a geosonic interface, which synchronizes the brain and body to planetary rhythms that we attend to even—indeed, *precisely*—in our sleep. This entails a shift toward nonconscious cognition (following N. Katherine Hayles) and a renewed conceptualization of what Catherine Malabou calls neuroplasticity, the brain's co-evolution with the body and the lived (though not necessarily consciously experienced) environment. The central artwork I examine in the chapter is Max Richter's *Sleep*, an 8-hour, fully notated piece of chamber music meant to be performed overnight for a sleeping audience. I listen critically to *Sleep* as a soundscape, transducing the geosonic and biopolitical forces that conduce and constrain how (some) bodies sleep on earth in the twenty-first century.

Chapter 6 makes the counterintuitive argument that Sun Ra—the prolific proto-Afrofuturist composer and bandleader who claimed he was from outer space—was actually a kind of geologist. I approach Ra's poetic and musical practice as a critical provocation to rethink terrestrial bias, exploring how his intergalactic music is grounded in a retuning of the technical and racial politics of listening on earth. According to the aforementioned Pythagorean music of the spheres (an ancient esotericism very much present to a range of mid-century European and African American composers, including Ra), music, sound, and implicitly, race have historically granted or denied epistemological access to the earth and the cosmos. In contrast to a dominant Western musical tradition that universalizes musical space (while historically displacing African and Afrodiasporic peoples from the cosmic-colonial soundscape), Ra posits a counter-mythological soundscape of real and imagined Black peoples, and I argue that earth is the material-metaphorical place in space which catalyzes his vision and practice. I focus on the rewording of everyday language in Ra's poetry as a transducer of utopian thought, and my close readings flow into an analysis of his reverberant compositional and audio techniques in his recordings of the mid-1960s.

The choice of material in *Geosonics* is deliberately eclectic and heterogeneous, characteristic of what geologists call an unconformity, as I explain in the

Afterword. Each chapter stems from my underlying conception of sound's multiplicity in/as listening, and is directed towards a single overarching critical goal: to listen through soundscapes as a way of thinking with the earth, and in doing so, to "reframe and deepen the discussions, both historical and philosophical [and media-theoretical], around the relationship between geology and humanity" (Rider and Harris 2023: 3). *Geosonics* is a critical experiment in attending to how sound shapes our thoughts; how sound shapes the earth; and how the earth, in turn, shapes our thoughts via its reverberations in our media and imagination (not to mention in new materialist, post-human, and critical-theoretical approaches to media and environment).

Geosonics is not an environmentalist book, though I intend it to open up new conversations among activists, earth scientists, and media and sound scholars, re-casting the earth not just as an object of study and a site of activism, but as a process of (aesthetic) mediation, and therefore, a shareable imaginary. In that sense, *Geosonics* is "pre-activist" (Jue 2020: 33), a book that wants to tune geo-scientists into the imaginary and everyday earth humming amongst the rocks; and to raise curiosity among humanists, artists, and activists for the inevitable, surprising ways that the earth resounds in the human imagination.

1

From Geosonics to Geosonicks

Part I. Audible Earths

The aim of this book is to examine, from a media-cultural and sound studies perspective, how people make the earth audible; and conceive of listening as an embodied experience and a technocultural act on a planetary scale. Despite the promise of the title, the earth is *not* the protagonist of *Geosonics* and aside from a brief cameo in the Introduction, barely makes an appearance in the following pages. But before the discussion inevitably shifts to questions of sound, materiality, and media as technical infrastructures of the listening imagination, let me consider two artworks where the earth might be said, rightly or not, to have its say.

Doug Aitken's *Sonic Pavilion* (2009) is an architectural and sound installation at Inhotim, a large cultural foundation and nature preserve (founded by a mining magnate), situated in the Brazilian rainforest. The installation consists of a hole, one foot in diameter and 700 feet deep, drilled into the earth and lined with concrete. Microphones and accelerometers are suspended at various depths inside the hole, and the sounds they pick up are mixed and piped into speakers arranged throughout a spacious glass pavilion at the surface, with the hole at its center. Visitors can look through the pavilion's curiously frosted glass windows (which are designed to blur the periphery and render clearly visible only what is directly in front of the viewer's gaze), and listen to what Aitken's website describes as "the sound of the earth turning and the tectonic plates shifting."[1] The pavilion is also an open-air structure, and the reverberations of the deep-earth sound merge with the lush sounds of the surrounding environment (bird calls, wind in the trees, rainfall, etc.) and the panoramic views to "create a living artwork." Through the sounds of the *Pavilion* we can experience the inorganicity of soil and rock alongside the teeming vitality of the flora and fauna it sustains. It purports to be a kind of ultimate ambient music, not just of the background, but the underground, too: a terrestrial and telluric soundscape.

But are we really hearing the sound of the earth's rotation? It is true that accelerometers measure minute changes in gravity and can thus detect planetary rotation. But there are accelerometers in your smartphone, tablet or wearable device that can be programmed to do the same thing without the hassle of subterranean installation. And the sound of plate tectonics? A 700-foot concrete tube is surely a wondrous conductor of sounds and vibrations. But seismic activity, strictly speaking, describes the internal movement of energy through rock. The air inside the column where the microphones are mounted is *exterior* to seismic movement. Seismic vibrations also propagate at frequencies well below the lower threshold of human hearing. So the question of how exactly the microphones use air to transmit the inaudible, elastic vibrations of rock is silenced (albeit pleasantly) in the musical ambience of the "living artwork."

Despite a certain literal-mindedness in its conception and the technical wrinkles that accompany it, I like to think of *Sonic Pavilion* as an ironic comment on the mining industry, a kind of reverse extraction in which deep earth boring techniques are used to redeposit the valuable metals contained in microphones and electronics into the earth (in much the same way that a mining magnate might redirect some of his capital to preserve nature, promote the arts, and so on). However, the seductive immediacy of a microphonic connection to nature and the earth has a surprisingly illustrious history in twentieth-century avant-garde music and the arts, most prominently in the work and writings of John Cage. Cage's "Future of Music: Credo" calls for "means for amplifying small sounds" as fundamental to new music ([1937] 2011: 25). By 1962, microphones had become the technical basis underlying Cage's famous *0'00"* composition (which I discuss in Chapter 3), in which sounds of everyday actions are amplified by contact microphones to demonstrate that "everything we do is music, or can become musical through the use of microphones."[2] Cage even theorized listening to the molecular vibrations of everyday objects, like an ashtray, which, if placed in an anechoic chamber and suitably miked, could reveal "the meaning of nature through the music of objects."[3] Cage's aesthetic aim was to direct attention away from human-centered actions (like composing music) to the aleatory happenings and indeterminate processes that fill everyday and nonhuman worlds with musical meaningfulness. Yet, as Douglas Kahn has argued, the expansion of the object of attention does not necessarily entail a change in attention itself: "Cage explicitly sought to subvert tactics based in human-centeredness, yet all he did was shift the center from one of utterance to audition" (1999: 197). Cagean listeners may no longer listen to a determinate creative act, but they do listen (to any and all sounds) with the same expectations of music and meaning as if they were in a classical concert hall. The microphone facilitates a disavowal of the technical mediations of the installation and thereby preserves an uncritical and antiquated mode of listening. In the case of the *Sonic Pavilion*, then, we would have to admit that the primary thing we listen to is not the earth, but the sound of microphones amplifying the air inside a giant, subterranean concrete tube. But rather than contemplate this truly remarkable set-up for what it is, we have to listen past it to hear "the sound of the earth." When it comes to listening, the more we ignore the technical infrastructure, the more we embrace a false, or at least rhetorically heightened, earthiness.

What does *Sonic Pavilion* actually sound like? Seth Kim-Cohen, who visited the site, writes:

> The sound itself is nothing special: Only the suggestion of its source elicits our attention and grants it meaning. Of course, this is always the case. Meaning does not simply inhere within the in-itself, regardless of whether it is the thing-in-itself or sound-in-itself. Meaning is only ever produced by the friction between things. Like every medium sound derives its meaning from context, from intertextuality, from the play of difference in its conceptual and material strata. It is the worldly, rather than the earthly, that presents the possibility of meaning.
>
> Kim-Cohen 2009b: 100

The core of this comment resonates with Kim-Cohen's larger thesis in *In the Blink of an Ear: Towards a Non-Cochlear Sonic Art* (2009a), where he argues that the Cagean rhetoric of "sounds themselves" is a conceptual dead-end that prevents sound art from dealing critically with conceptual and material questions. I would fully agree with Kim-Cohen's critique of the *Sonic Pavilion*, which argues that there is no essential or immanent earthiness out there, and that there is no clever or "correct" way to place a microphone that could be said to amplify the sound of the earth as such. It would seem that mediation of any kind would land us firmly in the *world* of meaning, but never on the literal, geologic *earth*.

The seeming opposition, if not outright exclusion, of world and earth is a reminder of another work of art that claims a kind of direct access to the sound of the earth. It is the painting of peasant shoes by Vincent van Gogh as described by Martin Heidegger in the essay, "The Origin of the Art Work" ([1936] 1977). For Heidegger works of art are unique in their ontological ability to connect the *world* where (human) beings live out their lives to the otherwise insensible *earth*, which Heidegger conceives of as the world's ontological substrate or periphery. The way that van Gogh depicts the peasant shoes affords an ontological revelation of the earth in precisely that humble interface of human uprightness and the earthly horizontal: the shoe. Even though we are talking about shoes and a painting, it is sound that carries the revelatory force of the earth. Heidegger writes:

> From the dark opening of the worn insides of the shoes the toilsome tread of the worker stares forth. In the stiffly rugged heaviness of the shoes there is the accumulated tenacity of her slow trudge through the far-spreading and ever-uniform furrows of the field swept by a raw wind. On the leather lie the dampness and richness of the soil. Under the soles stretches the loneliness of the field-path as evening falls. *In the shoes vibrates the silent call of the earth*, its quiet gift of the ripening grain and its unexplained self-refusal in the fallow desolation of the wintry field.
>
> 159, emphasis added

Instead of a microphone, it is a painting of shoes that transmits the vibrations, the call of the earth. Heidegger, like Aitken, downplays multiple levels of mediation to make an essentialist sonic connection to the earth. He writes about leather, wind, and soil, but strictly speaking, the object of his contemplation is oil on canvas. He personifies a peasant woman behind the shoes and only understands the weathering of the (pictorial) leather through the fiction of her imagined life. Finally, the alleged ontological depth of the (painting of) shoes only emerges in Heidegger's uncharacteristically vivid and descriptive language, including the mixed metaphor of shoes vibrating, calling out silently.[4] The more the shoes are conceptually and rhetorically framed and mediated, the more immediately the aesthetic connection to the earth seems to take effect. If it is not a technical instrument like a microphone, it can be the aesthetic imagination, working in relative silence, which manages to conjoin sound and the earth in some essential way.

The key conclusion to draw from the juxtaposition of these two artworks is that the earth, like anything else, may only be grasped through some combination of technical sensing, embodied sensation, and the reach of the imagination, all of which necessarily take shape within diverse cultural and discursive frameworks that change over time. As a technocultural construction, then, "earth" should always be read within scare quotes. The immense bulk of rock, water and air that surrounds us (or that we surround) only becomes intelligible through the means by which we (make) sense (of) it through our bodies, instruments, and imagination. Any introductory course in cross-cultural anthropology, religion, or the history of science will yield up a number of different earths depending on language, culture, technology, geography, politics, etc. Even though I have a particular earth in mind at the time of this writing (predominantly the Anthropocene earth of early twenty-first-century extractive capitalism), I recognize that the earth I am describing is produced in and through culturally contingent ways of knowing, (not) the least of which is my own writing. Throughout this book, I avoid essentialist claims about the earth, preferring instead to dwell on the technical and aesthetic set-ups that sustain such claims, not for what they reveal about the earth, but what they indicate about geo-technocultural relations at a given place and time.

The analytical work of *Geosonics* traces a general movement, across a range of musical, arts, and technoscientific contexts, from the "fact" of the earth—as soundscape, epistemological frame, ground of immersive experience, and backdrop of critical thinking—toward a mediated notion of earth as a technical and imaginative artifact. One exception is the music and writings of Sun Ra (discussed in Chapter 6), worth mentioning here as a curious counterpart to the earthiness we see in Aitken and Heidegger. In an undated poem titled "Earth Is A Hole In Space," Ra inverts the premise of an artwork like *Sonic Pavilion* to a seemingly absurd extreme. Rather than dig an actual hole into the earth and devise technical means of listening to the sound of its hidden depths, Ra reconceives of the entire planet as already a hole. He riffs on the space age, popular-scientific notion of a black hole as an anti-material, vortical void, suggesting that earth is equally a void (if a material one) because "its gravitational pull is all inward / and constructed to the people upon / this planet." Earth becomes a place of "negative transfers" that interrupt the "continuous continuity" of outer space. The materiality (or "material vacuumatron") of earth is a denial of space, a "non-existent of the space principle [. . .] surrounded by the walls of space" (Sun Ra 2011: 18–19).

For Ra, the earth is already a giant hole walled in by a kind of planetary concrete, but the sounds that resonate within it do not provide a more profound experience of its geologic depths or its silent call. Such sounds would be, for Ra (as a mythico-critical thinker of history, race, politics, and science), the echoes of a fundamental misunderstanding of the planet and its relation to the cosmos, and would call not for listening, but for a radical re-conceptualization of sound and planet alike: "Planet earth needs a new idea / another way of approach." I discuss the complexity of Ra's poetics of planetary sound in Chapter 6, but for now it is crucial to note Ra's singularity in refusing to take the earth for granted, indeed, for refusing to take the earth for any *thing* at all. From my perspective in *Geosonics* and in terms of Ra's poetic refusal of the earth (i.e., his geo-so-nix), Sun Ra emerges as the ultimate critical geologist for whom the earth is already always a figment of language and a ground of utopian reimagining.

Part II. Sound as Earth Media Interface

Sound has a material basis as vibrations, as waves of mechanical energy propagating through a medium. This is the sound Aitken is after in the *Sonic Pavilion*. Sound is also a metaphor, literally: a carrying-across that joins two things by differentiating them. This is the kind of sound that, I suggest, Heidegger hears in Van Gogh's painting. Rather than try to separate the material from the metaphorical, I join many scholars working on/in sound in trying to conflate those two dimensions as fully as possible, to listen to the material and the metaphorical as one and the same sound.[5] Throughout *Geosonics*, I approach this duality within sound as a doubling of the temporality of listening. When we listen, we hear vibrations in the air, a temporally bound transmission of energy. Yet we also hear those vibrations taking shape within our imaginations (inflected by culture, discourse, technological paradigms, and, not least, the physiology of our embodied experience); and in that sense, in listening, we hear those same vibrations "again" as meaningful sounds. At the same time, meaningfulness and imagination precede and produce the moment when we actually tune into these or those vibrations and hear them as sounds in the first place. When we listen critically, as aware as possible of sound's dual status as matter and metaphor, we do not so much hear sounds as "always already" culturally constructed; rather we hear sounds as part of an *always again* process. We listen to a sound, and listen to ourselves listening, and so listen *again*, at the same time, to how the material and the metaphorical co-precede each other. For listeners (and "listeners," as I argue in Chapter 4, applies to deaf and hearing audiences), sounds play themselves back as themselves; they have a capacity for self-remediation that is inseparable from the processes by which they transmit energy and affect.[6]

As affective and self-affective entities, sounds have a special status in recent new materialist philosophies and ecologies. Graham Harman, for example, has founded an influential philosophy of objects based on Heidegger's conception of equipment from *Being and Time*, which divides objects into the purposive, apparent, ready-to-handness of tools, on the one hand; and the "subterranean," "unearthly," and "otherworldly" ontological depth (i.e., tool-being) of objects themselves on the other (2002: 2). Objects are marked by an ambivalence or ambiguity that is ultimately grasped as an inner "reversal between the withdrawn tool-being of the world and its present-at-hand fragments" (26). Every object is, in a sense, two objects: it is what it appears as or what it can used for by some (presumably human) agent, but it is also what it cannot be grasped as or reduced to. In this way, Harman structures objects like sounds: they are feedback loops for themselves, oscillating between different dimensions or scales of presence and materiality, appearing and withdrawing in a kind of relational rhythm. Jane Bennett's concept of *Vibrant Matter* (2010) resonates in a similar way, drawing on Gilles Deleuze, rather than Heidegger. Bennett is interested in rethinking the anthropocentrism that has historically framed materialist philosophies. While acknowledging the "elusive recalcitrance" of matter, she also wants to describe the ways in which matter is suffused with (non)human qualities and capacities like vitality and agency (3). Her aim is to retune philosophy to attend to the vibrancy of matter, which both eludes our conceptual grasp, yet observably works in the world, albeit in unforeseen or unheard-of ways. From the perspective of sound, vibrant matter vibrates. The conceptual oscillation of things (between "recalcitrant" materiality and vibrant affectivity) makes all things literal/metaphorical oscillators, transducing conceptual or ontological dualities into physical, affective vibrations that demand listeners (e.g., philosophers and theorists) to redouble their attention, *always again*, to the material.

This is by no means an exhaustive survey of post- or neo-Deleuzian and Heideggerian trends in contemporary theory, but it is enough to suggest that relational ontologies and new materialisms share a deep (if largely implicit) affinity with sound. This affinity is made explicit in the engaging premise of Daniel K. L. Chua and Alexander Rehding's *Alien Listening* (2021), which uses a new materialist approach to the discipline of music theory, in order to develop nothing less than an Intergalactic Music Theory of Everything, that is, a flat ontology which "makes the universe audible as rhythm" (72). Building on a range of new materialist premises (including the aforementioned Bennett and Harman), Chua and Rehding describe the universe as a matter of frequencies, with all events/objects/processes manifesting themselves as frequencies, rhythms, and therefore, *repetitions*. Music becomes the key epistemological and theoretical tool for apprehending the universe's material structure because "[m]usic, as repetitive motion, does not merely move in time and resonate in space, but is materially embedded in these dimensions and can therefore model and disclose their properties" (70). The structure of any musical sound is a self-repeating, self-relating process, and even a single, isolated sound only emerges as such through its relational folding of the silence which precedes and follows it (74–5). In other words, in Chua and Rehding's universe, "[f]or something to exist, it happens twice. An event is never singular" (74). Their ontological argument about music complements my analytical approach to listening as an always-again process, attending to the putatively singular event of "a" sound relationally and repetitively through its material, technocultural, and imaginative stratifications of the sonic field.

Chua and Rehding's work recalls an earlier materialist ontology of sound, namely the "ontology of vibrational force" in Steve Goodman's *Sonic Warfare* (2012). However, for Goodman, the vibrational ontology has explicit political stakes because sound, besides disclosing the movements of affect and matter, is also a vector of social control and domination. Following in the footsteps of Henri Lefebvre's *Rhythmanalysis* and Gilles Deleuze and Félix Guattari's *War Machine*, the sonic-ontological field, for Goodman, is inseparable from and unthinkable without capitalism, racial politics, and military-industrial technologies. Related to this explicit politics (both as a cause and an effect) is Goodman's specific interest in the sound practices of Afrodiasporic musical cultures (in particular, dub, dancehall, and drum'n'bass) as sonic strategies of resistance in an ontological-political field. Considering Goodman's work (especially through the influence it bears of Kodwo Eshun's Afrofuturist manifesto *More Brilliant than the Sun* [1998]) alongside the exomusicology that Rehding and Chua outline, it is possible to say that the Afrodiasporic musical traditions of North America and the Caribbean already offer a rich *historical* example of what Chua and Rehding theorize *speculatively*, that is, listening and musicking from an alien, or in this case radically displaced and alienated, perspective. Through my discussion of Sun Ra in Chapter 6, I suggest that the speculative-theoretical project of outer space listening could be productively reconfigured as a historical study of Afrodiasporic and Afrofuturist musical cultures.

If sound is a process that self-mediates (for listeners) between its materiality and its meaning, then it is fair to say that a similar process is played out in the self-interfacing of things, which grants (philosophically minded) users limited access to and control over a materiality that otherwise works with an inscrutable logic of its own. In a certain way, we have come back to John Cage's ashtray, which—under the right technological circumstances—turned out not to be an object at all, but a process. It may well be that all objects are, like sounds, not fixed entities, but vibrant/vibratory processes, and new materialist and relational ontologies suggest that we retune thinking to the hidden processes underlying the object. But what are the philosophical equivalents of a microphone, or an anechoic chamber? How can we think and listen critically to the world of things without disavowing the technical set-up? And where on earth is *the earth* in all this talk of worlds, sounds and "subterranean" things?

In parallel to (and in some cases inspired by) the turn away from anthropocentrism in new materialist and object-oriented theories, media studies has already been rethinking the relations between objects and processes through reinvigorated conceptions of media, environment, and mediation. John Durham Peters reverses "the old idea that media are environments" into the claim that "environments are also media," which entails an examination of the dialectical ways that media tools and infrastructures are built into and out of natural environments (2015: 3). His concept of infrastructuralism fuses the legacy of (post)structuralist thought on language and culture with a media materialism that embraces the elemental (e.g., earthy, watery, airy) and infrastructural dimensions of media. For Peters, media theory needs to inquire into the "subterranean" depths of media, only not in the form of an anti-Kantian object-oriented ontology like Harman's, but as the historical study of media infrastructures, environments, and meanings. Peters writes: "The questions of how to define nature, humans and media are ultimately the same question. We know and use nature only through the artefacts we make—both out of nature and out of our own bodies—and these artefacts can enter into nature's own history" (51). Just as sound remediates its physical reality (as energy in a medium) into its equally valid realities as technical artefact and cultural experience, so do media, for Peters, exist and operate in (and reveal something about) nature and culture simultaneously. The philosophy of elemental media attends, always again, to the feedback loops of media artefacts, environments and the bodies and imaginations of the humans who make them.

The idea that media "enter into nature's own history" is inverted into the jumping-off point for Jussi Parikka's *A Geology of Media* (2015), which theorizes how "nature's own history" (i.e., geology) enters into contemporary media arts and artefacts. Parikka conceives of media on a geologic timeline and measures its effects on geophysical (and geopolitical) scales: "Geology becomes a way to investigate [the] materiality of the technological media world. It becomes a conceptual trajectory, a creative intervention to the cultural history of the contemporary" (4). Parikka is certainly inspired by aforementioned speculative realist and new materialist approaches, but I would argue his geologic take on media has more in common with Peters: geologic media pose not (just) speculative, but concrete questions about the historical and material entanglement of technology and nature. The "subterranean" ontological depth of an object like, say, a cell phone, becomes, in the media-geologic framework, literally subterranean (i.e., a question of the deep-time formation of rare minerals; their mining and manufacture; the carbon footprints of their use in the circuitry of the phone; and their after-life as e-waste in a landfill), showing the continuity between everyday media objects, geologic time, the geophysics of digital infrastructures and the geopolitics of labor. Parikka is not interested in uncovering the hidden affordances and agencies of, say, coltan ore, and celebrating a "life of metal";[7] rather the geologic angle opens a critical perspective on the geopolitics of resource extraction, manufacturing, energy consumption, and disposal that infrastructurally mediate how we use media. Media use is also earth use, and whenever we connect to a device, we also connect, in specific ways, to the earth, soldering the world-making capacities of media artefacts to the real and imaginative geology of the planet.

Media concepts, like Peters' infrastructuralism and Parikka's geology of media, thus pick up where more new materialisms and ontologies stop short, transforming philosophical speculation about vibratory and subterranean ontologies into technocultural questions about sensing and recording instruments installed at particular places and times on/in the earth; and about culturally contingent ways of knowing, experiencing, and imagining the data such instruments provide. The implicit sonicity of being(s) and matter as they oscillate, resonate and vibrate in contemporary theory may be listened to, always again, in a geologically minded theory of media. Geosonics, then, understands the mediation between matter and being(s) as sound, i.e., as a material and metaphorical energy that human instruments, bodies and imagination transduce into culture and experience. Such transductions are grounded literally/metaphorically in/on the earth, not as depth or background, but as an element that is immanently implicated in mediation itself. From the perspective of sound, mediation is always again geologic, just as the geologic is always (re)mediated.

Part III. Escaping the Soundscape

After all, the earth itself has always been the ultimate immersive environment.
Brian Massumi, *Parables for the Virtual* (2002: 142)

Murray Schafer's notion of soundscape, as I discuss in the Introduction, has been instrumental in conceiving of sound and listening as descriptive and analytical tools for understanding the (historical) relations between humans, technology, and the natural environment. But as Emily Thompson has elaborated, the soundscape is more of a mediation than Schafer might want to recognize: "Like a landscape, a soundscape is simultaneously a physical environment and a way of perceiving that environment; it is both a world and culture constructed to make sense of that world" ([2002] 2012: 117). If the soundscape only emerges insofar as it is perceived in cultured acts of listening, then it is fair to say that the soundscape is more of a filter than an index of the environment. For every sound heard in the soundscape (whether perceived as sonorous or noisy), there would necessarily be other sounds that go unheard, that are effectively excluded from each culturally and technologically contingent act of listening. This is less a problem of "sound-" than of "-scape," a Germanic suffix that enters the English language precisely through the word "landscape" and is tied etymologically to notions of creating, controlling, and shaping. A soundscape is already always a shaped, cultured sonic "view" that scales down the sonic environment to fit the auditory imagination.[8]

Thompson tracks the evolution of the modern American soundscape via new audio technologies in the early twentieth century, noting that "[a]s scientists and engineers engaged increasingly with electrical representations of acoustical phenomena, sounds became indistinguishable from the circuits that produced them" (118). Rather than serve as a means of capturing and representing a soundscape, new audio technologies (e.g., microphones and loudspeakers) and their electrical infrastructures could effectively take the place of the soundscape, and the electrical-technical criteria for evaluating the quality or experience of sound as *signal* became the de facto criteria for listening to the soundscape as a whole: "The desire for clear, controlled, signal-like sound became pervasive, and anything that interfered with this goal was now engineered out of existence" (ibid.). The soundscape, in other words, was no longer an environment to be listened to (with varying degrees of culturally specific selectivity), but was to be actively crafted by engineers listening across the physiology of the human ear and the physics of electrical signals. Thompson's argument that the modern soundscape be understood within a generalized technocultural framework of a modernity bent on the elimination of time and space reinforces the idea that soundscapes exclude as much as they include. In a perverse way, the engineering problem of noise elimination and the acoustic-ecological task of noise abatement are both instances of how the notion of soundscape eliminates or denies aspects of the experience of the sonic environment precisely in avowing to listen to it.

No longer "outside" in nature or environment, the modern engineered soundscape retreats into a seeming interiority of circuits and signals. Yet it might also be possible to listen through or across those technological instruments and hear the earth in which both instruments and listening bodies are embedded. Composer Pauline Oliveros has developed the concept of sonosphere as a kind of counterpart to the narrow conceptual bandwidth of soundscape ([2011] 2017). The sonosphere, a neologism analogous to words like biosphere or technosphere, is:

> the sonorous or sonic envelope of the earth created by all the vibrations set in motion by natural or technological forces that travel through earth from its core to beyond earth, air, fire and water as waves and phonons to receivers. Receivers are humans, all creatures perceiving and using earth bio and technological systems. Vibrations within the range of hearing may be processed consciously or unconsciously; vibrations beyond the range of the human ear are nevertheless received by the body and processed unconsciously or by other inhabitants of the earth and beyond.
>
> (115)

Expanding, perhaps exploding the notion of soundscape, the sonosphere encompasses a range of biological perspectives beyond the human and considers a planetary sonic field that extends from the core of the earth to the outer reaches of the atmosphere. In the sonosphere, "[a]ll cells of the earth and body vibrate" (113), and those vibrations—whether acoustic or electromagnetic, cellular or seismic, animal or machine, conscious or unconscious—draw listeners into a kind of complex vibratory unity with/on/of the earth. If the soundscape expels the earth from the practices and technologies that purport to listen to it, the sonosphere takes the earth in its totality (including all the technological and biological entities on it) as its object of listening. Yet here we seem to encounter the same problematic of the soundscape in inverse form: if our listening does not actively shape what we listen to, how can we make distinctions within the vast vibratory field in which we are immersed? How do we establish the spatial and temporal intervals within a vibratory unity, in order for sound to *be* sound (literally, as propagation across space and time) and be heard as sound? What are the specificity and place of the listener in the sonosphere? Oliveros tends toward the mystical in this regard,[9] but the paradox remains that sonic environments involve differentiation as much as immersion and that "[i]n the experience of sonorous immersion, one is on the outside of what surrounds one; one is sheltered in a space which one nevertheless oneself suffuses" (Connor [2005] 2011: 134). Is there a way to listen to the sonic environment of the earth that acknowledges how listening itself actively shapes what it hears, but can somehow attend to the outside of its techniques and tools, sounding, at the same time, what listening necessarily occludes?

My answer to this question is a transductive model of listening that understands how listening does not take place (in a pre-established soundscape or a primordial sonosphere), but rather *makes* place by etching out a mediated, mutually constitutive set of relations between listener, instrument, object, and environment. Stefan Helmreich has argued for such a concept of transduction at work precisely in an environment of literal (and ostensible) immersion (2007). He offers an ethnographic account of a dive on a deep-sea submarine and attends to the rhetoric of immersion (as immediacy, presence, merging of inside/outside) from the auditory perspective of the control and communication systems on board the submarine: "It takes techniques and cultural translation to carve a soundscape for humans out of the subaqueous milieu, to endow submarine space with sonic distance and depth, to create immersive space" (624). What Helmreich calls the submarine's "pinging sonarscape"—the pings detected from submerged transponders at fixed points on the sea floor and other sonar data—mixes with the air-pocketed acoustics of the vessel itself (wherein are heard the radio chatter between the submarine pilot and the surface vessel, not to mention the rock music from the pilot's MP3 player) and coalesces into an intimate auditory experience of immersion: "Submerging into the ocean almost seamlessly merges with a sense of submerging into sound—and into a distinctively watery soundscape" (621).

Immersion thus conceals a cyborgian experience of transduction. Listened to transductively, the seeming immediacy of the sub-marine soundscape becomes for Helmreich a highly mediated experience of boundaries, networks and communication systems that locate the cyborg body of the submersible, direct its outer movements and regulate its inner environment. Immersion in a soundscape does not happen just because one happens to be immersed in water: immersion is a technocultural experience that is transduced across a particular ocean environment (measured by depth, temperature, ocean currents, etc.), the vessel's technical systems, and the bodies and behaviours of its human cargo. To experience (a soundscape of) immersion means to inhabit multiple physical, technical, geographic and cultural locations simultaneously: to be (inside) a human body; inside a submarine chamber; inside the circuits of its cybernetic systems; immersed in particular location in the ocean environment; and finally immersed in a cultural rhetoric of "immersion" that structures how one imagines (and does ethnography about) immersion in the first place.

Helmreich effectively adapts the concept of soundscape into a self-reflexive model of transduction that distributes the experience of a soundscape across layers of material, technical and cultural systems. A soundscape is thus more and less than a sonic environment: *more* in the sense that it is a multi-scalar ensemble of environment, instrumental infrastructure, cultural rhetorics and embodied practices of listening; *less* in the sense that a transduced immersive soundscape only reminds us of the partial, mediated and culturally constructed ways we can listen to *any* sonic environment. What about the earth, the immersive, vibratory field that is somehow still "out there" beyond or beneath the technocultural interiority of the soundscape? To answer this question, we might change the emphasis of Helmreich's argument: the soundscape is not only composed of multiple technocultural transductions, but the soundscape is also a transducer itself.

The submarine on which Helmreich is conducting his research is itself an oceanographic research vessel tasked with mapping an unmapped portion of the sea floor (the Mothra Hydrothermal Vent Field in the Canadian Pacific). As the vessel moves out of the mapped territory and into the as-yet unmapped zone, the on-board scientist announces to Helmreich, "'We are merging with our data'" (630). Helmreich notes the potential here for a kind of Borgesian merging of the map with the territory, where one could imagine the transductive experience might open onto some kind of genuine immersion after all, in the "real" of the "un-scaped" unknown. Yet in the next paragraph, Helmreich abruptly displaces the scene of this merging/immersion from the deep-sea threshold to the water's surface:

> A couple of days later, at a science meeting on *Atlantis* [the surface vessel], Delaney [the scientist] enacts his sense of merging corporeally. As he reviews the topography of Mothra, he directs a postdoc—the person who painstakingly created the final graphic—to pan and tilt a three-dimensional computerized map, projected on a video screen. He moves his body like a conductor and even

> says, "Music please," embodying the orchestrating, directing relation of professor to postdoc so characteristic of the natural sciences. In this synesthetic dance, his body fuses with the map; he merges with the data.
>
> (630)

Immersion happens paradoxically at a temporal and spatial remove. What seems at first like the ground-zero moment of a transductive process, that off-grid encounter that will summon a topography out of the unknown and into being, generate a territory out of chaos etc., in fact emerges as yet another proleptic encoding of data that will only enact further transductions: the "painstaking" visualization by the unnamed postdoc; the embodied performance of the data by the scientist; and the overall re-gridding, re-scaping of the earth as one more puzzle piece is slotted into the oceanographic map by the scientific community. Merging with the data and immersion on the uncharted ocean floor do not happen *in situ*, but rather days later when a scientist engages verbally and bodily with projected images of a data visualization. The underwater soundscape aboard the submarine vessel is more of a sound*ing*scape, a transducer in its own right that transforms sonographic data into visualized data into an embodied performance of scientific knowledge. The earth is heard in—or emerges from—the soundscape, but only belatedly and insofar as it is transduced by the datic and bodily transformations that map out what it was in the first place. Put another way, the soundscape is less an environment than an instrument, which transduces geology into a communicative embodied human experience across space and time.

Helmreich's take on transduction amounts to a deconstruction of the rhetoric of immersion, yet with an emphatically materialist twist. Considered transductively, immersion does not dissolve into an ocean of metaphor and textuality from which there is no outside; rather, immersion is a (deconstructible rhetoric of) *presence-with-an-infrastructure* that conjoins whatever cultural fictions we may ascribe to it with material instruments and bodies, transducing an environment out of a technocultural feedback loop. Following John Durham Peter's concept of infrastructuralism, we might think of transductive analysis not as deconstruction, but infrastruction: how an experience of (auditory) presence is buried in a technological infrastructure that disappears from our senses precisely in the moment when we experience "unmediated" immersion in an environment. The analytic of transduction thus helps both to conceptualize and listen to the outside of the soundscape, to the tools and spaces that are displaced and deferred by techniques of listening to the sonic environment.

Helmreich has elsewhere pointed out that the concept of transduction has played precisely this role of uniting the material-technical with the cultural-semiotic in the field of sound studies (2015). Echoing Jonathan Sterne's work on transduction in *The Audible Past* (which I discuss in Chapter 2), Helmreich suggests that transduction serves as the conceptual intersection of science and technology studies on the one hand, and media and cultural studies on the other, all under the auspices of the study of sound (Helmreich 2015: 223). The world of culture, lived experience, and meaning is inseparable from the earthy infrastructure of tools, materials and, ultimately, geology itself. Analytically, transduction audits the ways that the world of the soundscape is embedded in the earth, but experientially, transduction *is* the mediation that makes the earth disappear after all into the seeming immersion of our auditory experience.

For Helmreich, this means that subjects, objects, media, and environments do not precede, but rather are constituted by the transductive processes they animate.[10] Even the seeming indexicality of mapping (as in Helmreich's dive) must pass through the relay of a transductive fiction. The presumed fixity of the earth—not just on a map, but literally "on the ground"—is less a categorical given than a state of affairs that must be continually (re)established by transductive listening. The earth (qua that parcel of hydrothermal vent to be mapped), we might say, comes into being through a transductive process of *enlistenment*. Tim Ingold has written of the phenomenology of weather and of sound, and, playing on the phenomenological concept of embodiment, describes the body's enwindment and ensoundment in breath. He stresses that as the living body breathes and participates in the airy atmosphere, it does not so much embody the wind as become enwinded. The sounds of breathing (qua enwindment) draw the body outwards into the sonic environment, rather than internalize a fixed soundscape, and the body is thus ensounded as it breathes (Ingold 13). Considering this reversal of the soundscape transductively, soundscapes are transducers that enlisten the earth and enearth listeners. If the earth really is the ultimate immersive (auditory) environment (as the epigraph from Massumi suggests, albeit in a different context), this means that our ways of attending to that environment must follow a circular or spiralling logic of transduction in which the earth is perpetually unsettled and resettled in and by sound:

> Sound flows, as wind blows, along irregular, winding paths, and the places it describes are like eddies, framed by a circular movement *around* rather than a fixed location *within*. To follow sound, that is, to listen, is to wander the same paths. Attentive listening, as opposed to passive hearing, surely entails the very

> opposite of emplacement. We may in practice be anchored to the ground, but it is not sound that provides the anchor.
>
> Ingold 12

If we substitute "transductive" for "attentive" listening, it becomes clear that immersive concepts like soundscapes (and even immersive audio technologies like sonar) do not so much locate bodies in space as they distribute bodies across time and space, and those temporal and spatial coordinates circularly coalesce into the sonic environments that they were in the first place. Soundscapes are not acoustic or vibratory emanations of environments, but rather passages that reveal/conceal the infrastructures by which emplacement and embodiment seem to coincide in our experience.

Part IV. Interfaces Again: Geosonicks

Geosonics, as the mediation between sonic elements (i.e., bodies, instruments and the earth), implies transductive listening, that is, attending always again to the material embedded in the cultural, and vice versa. This entails listening across a particular transductive infrastructure at a particular time. The geosonic earth, then, is special kind of technocultural construction that is always again enlistened, that is, produced in and by technical acts of listening. Likewise, from a geosonic perspective, the experiences of bodies are not only embodied in relation to a world, but enearthed, that is, they participate in processes, relations and infrastructures of mediation that are earthly through and through.

In this regard, I part company with the philosophical and phenomenological conceptualizations of transduction that have a particular currency in the processual materialisms I discuss above. For Maurice Merleau-Ponty, for example, transduction refers to the mutually constitutive processes of identification and differentiation between a self and a world (Hansen 2006). The mediation of the senses is a kind of structural gap that gives the body its interiority in transductive interplay with the exteriority of the world. For Gilbert Simondon, transduction names the process by which milieus or territories are ceaselessly generated and transformed by individuals who are themselves unfolding and becoming in the self-same, ontogenetic flux of the transductive process (1992). Merleau-Ponty presumes a world as a neutral backdrop, just as Simondon (and later Deleuze) presume an infinitely available, expandable supply of territory-generating matter.

In contrast, I see "world" as a limit to transduction and transductive thinking, rather than the medium or site where the process occurs. As Timothy Morton has argued in *Hyperobjects: Philosophy and Ecology after the End of the World* (2013), the philosophical concept of world undermines serious earthly and ecological conceptualization. For Morton, worlds cannot link the immanence and finitude of the earth to the being and experience of objects; worlds only perpetuate the asymmetrical reign of subjects over objects, risking blindness to the hyperobjects (like earth, climate, capitalism) that upset the figure/ground, subject/object, culture/nature dichotomies that frame the phenomenology of self/world, or of expansionist milieu: "What ecological thought must do, then, is unground the human by forcing it back onto the ground, which is to say standing on a gigantic object called *Earth* inside a gigantic entity called *biosphere*" (18). Following Morton, I replace the human/world dyad with body/earth, and attend to the ways that bodies are enearthed, that is, embedded not just in themselves, but in aesthetic experience of the (hyperobject called) earth.

The earthly (rather than worldly) aspect of geosonics can contribute to contemporary critical thought on related geo-scaled topics like anthropogenic global warming and extractive capitalism because it substitutes the ultimately binarizing conceptuality of immersion (in a system, in a world) with a critical concept of transduction that can tune into the scales, infrastructures, and affects that transduce planetary processes into our imagined and lived experience. *Geosonics* is not exactly an environmentalist or ecological study, though the concept takes as a significant point of departure the literal/metaphorical "end of the world," as well as the "daunting, indeed horrifying, coincidence of human history and terrestrial geology" that follows it in its wake (Morton 2013: 9). Expanding on Morton's move to counteract the "bad" Heideggerianism of, well, Heidegger, with the "good" Heideggerianism of object-oriented ontology and speculative realism, my aim is to listen through soundscapes as a way of thinking through the geologization of the human in the context of a critical media studies of sound.

Jacob Smith's *Eco-Sonic Media* (2016) approaches similar concerns around sound and media, but from a different angle. His approach is a "green-media archaeology" of sonic media inspired by ecocritical media studies that examine the environmental impact of media consumption, holding media studies accountable, methodologically, for the carbon footprint of its object of study. Key studies in this direction include Richard Maxwell and Toby Miller's *Greening the Media* (2012); Allison Carruth's "The Digital Cloud and Micropolitics of Energy"

(2014); Nicole Starosielski and Janet Walker's *Sustainable Media* (2016); and Sean Cubitt's *Finite Media* (2017). In stressing the imaginative and experiential dimension of mediation (even if that dimension is also always earthly, planetary, geologic), I depart from the methodologies (if not the concerns) of more ecologically minded media studies. The difference between the eco- and geosonic can be described by the difference between media and mediation invoked by Alexander Galloway (2012). Studying media means examining the symbolic and material affordances intrinsic to specific media artefacts, while the study of mediation looks at the practices and techniques that link users and artefacts to a socially inflected media system. Mediation is a question of interfaces and processes, not objects. Translating Galloway into ecological terms, the compression and consumption of earthly resources in and by (audio) media artefacts (i.e., eco-sonics) is distinct from the symbolic and material interfacing of (audio) media users with the material and symbolic earth (i.e., geosonics).

In that sense, geosonics conceives of the earth less as a hyperobject, and more like an interface. I have already suggested, via Peters and Parikka, that all media artefacts are earthly *and* interfaces with the earth, and that media (like sounds) refer us to cultural meanings and earthly materiality simultaneously. But the term "interface" here requires more nuance beyond its conventional meaning as intersection between technological device and user, or point of operational connection/synchronization between machines. One of Parikka's theoretical interlocutors is German media archaeologist Siegfried Zielinski, who developed the "deep time" approach to the history of media that informs Parikka's *Geology*. For Zielinski, media are "spaces of action for constructed attempts to connect what is separated" ([2002] 2006: 7), and the crucial problem of media is the interface (*Schnittstelle*), the place (*Stelle*) that marks the cutting and joining (*Schnitt*)[11] of human worlds, realities, and scales of experience with some kind of alterity that is, at different points on the media-archaeological timeline, imagined as the alter-interiority of machine worlds or the alter-exteriority of geologic time. Either way, media offer a "cut through the world, which enables it to be experienced" (32). Zielinski continues:

> [W]e swim in [media] like the fish in the ocean, it is essential for us, and for this reason it is ultimately inaccessible to us. All we can do is make certain cuts across it to gain operational access. These cuts can be defined as built constructs; in the case of media, as interfaces, devices, programs, technical systems, networks, and media forms of expression and realization, such as film, video, machine installations, books, or websites.
>
> (33)

There are, in a sense, two interfaces at work in this passage: the specific interfaces of media devices and media arts, and the general interface of any media device with the larger, unknowable totality on the other side of our experience. Parikka's take on that totality is to implicate the earth as the ultimate interface, in and by which all particular media devices are to be constructed, conceived of, and critiqued. Media always imply a technical and imaginative construction of the earth by virtue of the same earthly materiality that gives rise to them. But Zielinski stresses a paradoxical aspect to this self- or re-mediating process: the more media try to access or imagine the whole, the more cuts they make in it. Media, as interfaces, hack the erstwhile earth into pieces, building smaller media worlds out of the ever-increasing fragmentation of the whole (whatever that whole might be). Any discussion of the "world-making affordances of media" must also acknowledge that media work by breaking up the earth, one piece at a time, with each world they construct.[12]

It is in this sense of interface that I want to write geosonics again as geosonicks. As the *Oxford English Dictionary* will attest, "nick," like "sound," is a tricky English word—and here it bears remarking that dictionaries are important geosonic instruments insofar as they transduce and record the movements of meaning that accrue around a particular material entity (a sound and its textual form as a written word). In any case, to nick means to make a notch or groove in something, cutting into or chipping away at a surface either accidentally, or for the purposes of keeping score, or indicating ownership. It can mean to shape, or to deform. In the context of mining, to nick means to remove coal from a seam. The word can refer to a fortuitous or precise moment in time (in the nick of time); a favorable condition (in good nick); or a winning play (in dice, cards or the game of squash). But it can also mark how the fortuitous devolves into the appropriative: in slang, to nick means to steal, cheat, or defraud; to take into custody or under arrest.

There are three aspects of geosonicks that I want to stress with this portmanteau formation: (1) mediation of any kind (geosonic or otherwise) nicks away at the earth: it burrows and borrows, shapes and disfigures, measures and steals. Mediation may be productive, but its infrastructure is always extractive and appropriative. Its success is always a little like theft. (2) Mediation implies a permanent and necessary partiality or incompleteness. Like Zielinski's literal take on the interface as a cut, geosonicks implies a necessarily part-for-whole logic at work in mediation: we can only experience the "Whole Earth" in the fugitive pieces of interfaces we build into and out of it. (3) Geosonicks marks a moment in time, a particular technocultural juncture of listening and media interfaces corresponding to a specific configuration of geology, technocultural imagination, and geopolitics. In opposition to the Deleuzian notion of geophilosophy, in which the primordiality of the earth is encountered over and over in a perpetual refrain of territorializing flux, geosonicks aims to critically understand how earth is transduced into technology and experience, always contingently, in historically distinct and describable ways.[13] Geosonicks involves a timestamping of earth, bodies, and instruments by transductive listening.

2

The Sound Beneath Our Feet

Earthquakes and Ear Quakes

Part I. Listening Beneath the Soundscape

Any discussion of the earth, whether in a scientific, cultural-theoretical or everyday sense, is marked by a slippage: earth sometimes refers to a planetary totality spanning from the earth's core to its outermost atmosphere, including every thing and element in between. At other times, the term has a more local, terrestrial cast, implying the specificity of rock and soil, as opposed to water and air (not to mention organic life). But geology is the study of the long-duration interactions of earthly elements with watery and airy atmospheres, so invoking the earth from this perspective necessarily entails a continuum of earth-as-planet with earth-as-territory/-terrain. In a similar vein, I have set up the term geosonics to engage a multiplicity of technical and imaginary sonic constructions of the earth as: (1) a planetary or global field of sound; (2) the vibratory specificity of singular (geologic) sites and events; *and* (3) the technical and imaginative transductions that allow human listeners to inhabit corporeal, terrestrial, and planetary fields simultaneously.

Soundscapes are, as we have seen in Chapter 1, nominally tethered to specific sites on the earth, even as, medially, they swirl around in the fluid atmospheres of air and water, eardrums and instruments. Roughly in parallel with Schafer's work on soundscapes in the late 1960s and the development of acoustic ecology, the field of seismology developed tools and techniques for listening to the earth by detecting and displaying the seismic vibrations of earthquakes, underground nuclear testing, and other geophysical activity. If soundscapes have infrastructures that are both geological and technological, and if those infrastructures are, literally or metaphorically, buried underneath our embodied experiences of listening to them, then the sonics of seismicity explicitly shifts our listening to the technological and earthly outside or underside of the soundscape.

Indeed, as we will see in detail below, it is a curious short-circuit of the technocultural feedback loops of modern seismology that earthquakes were initially recorded, conceived of, and listened to as (musical) sounds, with clearly describable acoustic contours. In seismology's coupling of earthquakes with the human ear, there is a kind of echo of what Jonathan Sterne calls the "tympanic function" (2003). For Sterne the tympanic function of the human ear (i.e., the eardrum's status as a transducer of sound waves in the air) was crucial in the evolution of sound recording around 1900, serving not only as the mechanical basis of recording and playback technologies, but indeed as the conceptual underpinning for understanding what sound, hearing, and listening were at all. But that function very quickly "would take on a life of its own" as "the technologies [it gave rise to, e.g., telephone, phonograph, microphone, and radio] were gradually organized into media systems with their own distinctive industrial and cultural practices" (84). In a certain sense, seismology belongs to the technological afterlives of the tympanic function because, especially in the mid-twentieth century, it attempts to render the earth intelligible by recording it and playing it back as sound. In this chapter I sketch out how the epistemological stakes of seismology (i.e., what it can claim to know of the earth by studying how it vibrates) are folded into technical questions of sound and signal processing. The construction of the earth as an epistemological object has been and continues to be shaped by pre-existing ideas about sound, recording, and human listening. It is not unreasonable to assert that we know the earth through our eardrums. This assertion, at the same time, opens up the earth (qua seismological entity) as a field of activity for sound artists, composers, and musicians. The aesthetic reimagining of geophysical data and the (literal or metaphorical) artistic engagement with earthquakes can reveal, with varying degrees of critical awareness, the ongoing transductive processes that make up the ground beneath our feet.

Since the seismic vibrations of earthquakes usually have a frequency of 1Hz and rarely, if ever, move into the audible range,[1] they present us with an example of a sound(scape) that cannot be experienced through (a rhetoric of) immersion, but rather which foregrounds technical transductions and signal processing as part and parcel of its sonic material. To listen to an earthquake means to listen to a sound that can both claim to have a solely transductive existence as a phantom of circuits, scales, and ratios, and at the same time be heard as the vibration of the entire ground and infrastructure of sound itself (on earth). If Stefan Helmreich's analytic of transduction aims to "listen for that which we only usually hear"

(2007: 629), that is, to attend to the instruments and materialities that disappear into the experience of the soundscape, then to listen transductively to earthquakes is an ideal practice of listening *through* soundscapes.

Earthquakes—as transduced and imagined by scientists and technicians—do away with the notions of ambience, atmosphere and immersion that permeate the concept of soundscape. There is no background hum of "keynote" sounds against which "soundmarks" distinguish themselves in an acoustic version of the figure/ground relationship, nor is there a fluid ambience into which listening can dissolve or disperse. The earthquake presumes no environmental backdrop or container in which acoustic or vibratory events unfold; rather, the earthquake is *all* event, inside and out. John Durham Peters has written of seafaring vessels as media and metaphors, and notes that "[i]n an extreme situation [at sea] everything on a ship is cargo, including the ship itself" (2015: 104). Something of that logic applies to the extreme situation of earthquakes, where the figure *is* the ground, where the soundscape is all sound, no scape. In the likewise extreme situation of anthropogenic global warming, climate is less and less a diffuse background of our habitation, and increasingly an event unto itself. There is thus a kind of homology between our experience of sound(scapes), earth(quakes) and climate (crisis): we cannot make an absolute distinction between ourselves as subjects, the environments we are in, and that which we think we are listening to/standing on/living in because in moments of crisis the medium, the environment, the subject and the object are all (in) the same boat, as it were, drawn together into an infrastructural unity punctuated by extremity or catastrophe.

Although the concept of soundscape is intrinsically bound up with an ecological stance against noise pollution, Steven Connor has suggested how it may actually be mid-century avant-garde composers who have more presciently anticipated contemporary ecological concerns than Schafer's acoustic-ecological paradigm. His talk, "Strings in Air and Earth," shows how the airy possibilities of Cagean "sounds themselves" and the push toward open-ended, indeterminate and long-duration compositional styles developed alongside an exploration of the more "down to earth," densely textural sounds of György Ligeti, Giannis Xenakis, and Karlheinz Stockhausen. A piece like Ligeti's *Atmosphères* (1961), for example, "seems to involve some mutation or crisis in the imagination of air, which has for so long provided the metaphorical support for imagining the materiality of music" (Connor 2006: 6). Rather than conceiving of the air as a limitless space for the transmission and reverberation of sounds and soundscapes—a space where listening might expand infinitely as well in its potential attentiveness to any and all sounds as music, Connor suggests that the atmosphere of late twentieth century music is not just finite, but crowded, colonized, polluted: "[postwar m]usic is conceived, presented and experienced as clustering, congelation, swarming, aggregation" (4). He continues: "Diffusion implies a space into which sound, like any other waste product, can expand and slowly vanish. During the twentieth century, the air has gradually been finitized. For us, the air has lost its inviolability, its capacity to dissolve and diffuse any pollution. Itself finite, and therefore vulnerable to damage and corruption, the air no longer provides the promise of universal purification" (10). Music, via the composers Connor singles out, can both prefigure and respond to this crisis of the air with sonic practices that intensify, rather than bypass, those soundscapey thresholds when music aims to step out of the foreground and lay claim to a taken-for-granted, diffuse background. Although he does not use the term transduction, Connor has in mind precisely that concept when he calls for the "intensification and thickening of the transactions between foreground and background, signal and noise, hearing and listening" (11) as a means of sounding out the complexity of finite musical material previously taken for granted as atmospheric, immersive, infinite.

The earth's atmosphere, as Timothy Morton reminds us, "contains throughout its circumference a thin layer of radioactive material, deposited since 1945," not to mention all the spent hydrocarbons deposited since the invention of the combustion engine by James Watt in 1784 (2013: 4). Connor's account, via Morton, shows how music presciently conceives of atmosphere as Anthropocene: not an ethereal escape from the man-made world here on earth, but an element that is already populated with human traces. The musical air is, in effect, enearthed by pollution: it is neither a vast, empty space into which our waste products can somehow disappear, nor, metaphorically, an uncluttered musical ambience where composers and musicians can unproblematically discover alternatives to historically encrusted musical forms and listening habits. Consider the contrast between a piece like *Atmosphères* and a seminal work of ambient music, Brian Eno's *Music for Airports* (1978). While Ligeti's piece, via Connor, seems to stage a struggle between sound and its medium (the air itself), Eno's ambient approach simply suspends the crowds, noise and tension of airports, as well as the massively congested traffic and debris of the upper atmosphere itself.[2] The music—minimal, unpulsed, looping, ethereal—presumes there is an ambience somewhere out there and offers musical and affective cues for its sonic habitation, but the conceptual stakes of the piece do not grapple with the material and ecological dimensions of the site or situation it wants to address. Aesthetically, then, *Music for Airports* is both a Muzak-like, affective regulator and a concert-hall piece that foregoes the actual environment for a transcendent musical domain (an ambiguity I discuss further in music for sleeping in Chapter 5); but it does not engage music, listening and environment geosonically. It does not thicken the transactions (i.e., transductions) between what counts as foreground and background (as Connor has it), but rather smooths and rarefies the foreground into the background.[3] In this chapter, I am interested in how the underground fits literally and metaphorically into this problematic. If the underground (of modern seismology) is recorded, played back and apprehended as sound, then how do artists working with sound appropriate that ultimate sonic territory?

Part II. Earthquakes: Music to Sounds

As mentioned above, earthquake sounds are not sounds per se in that the seismic vibrations caused by earthquakes almost always propagate at frequencies below the threshold of human hearing. But earthquakes and sounds share a similar structure as waves, and seismologists have long exploited a kind of proto-sonicity of earthquakes through techniques of auditory display. A counterpart to data visualization, auditory display refers to the scaling and playback of complex data sets in audio formats as an analytical and pedagogical tool (Kramer 1994). Technicians usually make a distinction between audification and sonification. In audification, a data set is regarded as already "shaped" like a sound wave and can thus be played back without any other manipulation than tweaking the frequency range and time scale. This is how seismographic data is typically audio-displayed, according to geoscientist Chris Hayward: "[s]ounds transmitted through the air (acoustic waves) have similar physics to seismic vibrations transmitted through the earth (elastic waves). The physics is similar enough that mathematical models that describe sound transmission through gas are successfully used for seismic modeling" (Hayward 1994: 370–1). Sonification, by contrast, involves transposing data values onto a predetermined set of musical frequencies or sounds, as a way of making certain patterns audible, rather than visualizable. Sonification, not audification, would better characterize the cyborg interface I describe in Chapter 4, where Neil Harbisson "hears" color frequencies as a more or less arbitrary, but consistent (i.e., sonified) range of musical tones programmed into his prosthetic antenna.

The distinction between audification and sonification is deconstructible to an extent: audification, as in the audio seismograms I discuss below, implicitly or explicitly understands itself as listening directly to "raw" data, whereas sonification appears more as a translation across media formats, a mapping of data into sound for display purposes, but without a perceived intrinsic connection between recorded data and audified playback. The catch is that the "raw" data of seismology has already been cooked by pre-existing technocultural constructions of sound (as mechanical wave energy) and hearing (as tympanic transduction). When we listen to a highly composite technical artefact like an audio seismogram and imagine that we are hearing the earth quake in our loudspeakers or in our ears, we indulge in a transductive disavowal that bypasses the multiple technical and conceptual interfaces that comprise a sound in the first place, let alone the sound of an earthquake.

In the context of the seismological discourse I engage below, earthquake audification appears less as a transformative feat of data processing and more like the editing of a pre-existing sound. After the initial moment of transduction (in which seismic vibrations are registered as movement by seismometers or geophones and transcribed into seismograms), the data set of an earthquake is already like an acoustic sound with a clear pattern of attack, sustain and decay. Chris Hayward argues that earthquake audification is a unique fit between geologic and human scales and "can take advantage of the vast human experience in interpreting noises" (371). If the audification process preserves the shape and the structure of the seismic waves amidst the processing techniques necessary to transpose it to the audible range (e.g., time compression, frequency doubling, automatic gain control, etc.), then it will sound "like a recording of natural, environmental sounds." Hayward goes on to describe the initial attack of an earthquake as "two sharp sounds (reminiscent of a heavy door with a latch closing)" (398) and stresses that "a low pianolike sound must remain pianolike when shifted in frequency" (398) and that the overall process "must preserve the characteristic attack and decay of impulsive sounds as well as harmonic relationships" (384). For geoscientists like Hayward (and Florian Dombois, whom I discuss below), the geologic scale of the earthquake is surprisingly at home in the human ear. Hayward's descriptive metaphorical register ("reminiscent of a heavy door") blurs into a prescriptive mode of apprehending the data in the first place: the data is already "pianolike" and characterized by "harmonic relationships" *before* audification, and that technical process is understood as preserving, rather than embellishing, the inherent sonicity, even musicality of the earthquake.

For Hayward, earthquake audification has an immediate practical value for seismologists in carrying out routine diagnostic testing of their equipment. Non-earthquake-like sounds (caused by malfunctioning equipment) are much more readily and efficiently heard through audification than read off of the squiggles of seismograms. Indeed, the first scientific paper on earthquake audification hinged the scientific value of the process on another kind of geosonic pitch discrimination, that is, on making accurate distinctions between the seismograms caused by earthquakes versus underground nuclear explosions. Written at the height of the Cold War, when global nuclear test monitoring was a matter of keen national and military-industrial interest, Sheridan Dauster Speeth's "Seismometer Sounds" (1961) demonstrated that listeners (in this case, a group composed of Bell Labs technicians and some local high school students) could be rapidly trained to distinguish with ninety percent accuracy the sound of an earthquake from that of an underground nuclear bomb. Listeners would hear recordings juxtaposing earthquakes and nuclear explosions in a regular pattern. After a few listening sessions, they could then accurately discriminate between quakes and explosions when listening to a random sequence of audio seismograms. In a section of the paper titled, "Transformation of Explosive Sounds to Earthquake Sounds," Speeth writes that "[t]he subjective distinction established by the experiment described above must rest ultimately, of course, upon physical differences between the classes of events" (912). In other words, the specific seismicity of an underground nuclear explosion (about which relatively little was actually known to seismologists) could be hypothesized as fundamentally different from that of an earthquake only through audification, that is, through "subjective distinction." The differences were not otherwise available to scientists using visual or computational tools alone.

Speeth goes on to suggest that adding filters to the earthquake sounds to make them sound more like explosions would be a way of investigating the seismic patterning of the explosions themselves. Tweaking audio data into a simulation of an explosion for a human listener could be used to deduce an understanding of seismic events. Altering the audified model, then, could be effectively treated like a new data set, as if it came from underground. Epistemologically, this suggests that planetary structure is revealed in and by the human ear. Just as the sonar data of Helmreich's dive to the sea floor did not effectively generate a map until it was re-transduced offsite by human embodiment and data visualization (as I discuss in Chapter 1), so does the transductive logic of seismic audification suggest that underground explosions become not only identifiable, but also understandable as such when they are simulated before discriminating human ears.

There are three geosonic aspects of Speeth's paper to stress here: (1) we are invited to think of the earth as, all at once, an artist, composer, performer, instrument, and body. The earth composes or performs sounds which resonate in its own body like the resonant body of an acoustic instrument; (2) we think of seismologists as recording engineers who are shaping and aestheticizing the material in very specific ways according to preconceived notions about how it is supposed to sound, indeed, of how sound is supposed to sound. Just as a recording engineer might use certain kinds of microphones in a certain spatial disposition to simulate an ideal (or an actual or historical) acoustic space in recording, so do seismologists manipulate data to conform to sonic conventions even as they evaluate them; and (3) the transduction of these particular subterranean sounds (especially the *timbral* differences between quake and bomb seismograms) is unthinkable without Cold War-era military-industrial research. The geosonics of earthquakes *c.* 1961 is also a geopolitics of earthquakes, and as geopolitical realities evolve, so do the very sounds of the earth qua earthquakes.

Before audio seismograms became a data display tool of serious scientific interest, they were a kind of musical curiosity among seismologists and even found their way onto an experimental LP put out by Cook Laboratories titled *Out of This World* (Benioff and Morgan 1953).[4] On the record, Caltech seismologist, and inventor of the modern seismograph, Hugo Benioff presents and narrates a series of earthquake sounds recorded from a seismological center in Pasadena, California. His narration, interjected dramatically amidst obscure rumbling, crunching and booming sounds, begins, "This is a record of actual earthquakes." The word "record" might refer simultaneously to the archiving of seismic events, the recording output of the seismograph and the vinyl disc rotating on the listener's turntable. It almost seems, via Benioff, that earthquakes were conceived of as musical recordings even before they were conceived of as sounds.[5] Benioff's narration goes on to give a lesson in geosonic scaling. First, he explains how time-compression transposes earthquake waves into the audible range, and how factors like playback speed and the distance of the quake from the recording station in Pasadena affect the timbre of the seismogram. While

Benioff does not explicitly use the word timbre, his descriptions of earthquake sounds under different time-compression or distance factors (e.g., "the brittle, bone-crushing texture of these sounds" or "a sort of feathery tone about an octave below middle C on the piano") invite our musical, rather than scientific listening to the geosonic material. The juxtaposition of different earthquakes by distance factors (a local quake off the coast of San Clemente; a more distant quake in Hawaii; a third quake on the other side of the planet, south of Madagascar) lends itself to what Benioff calls "a sort of perspective of the earth's acoustics," and the aftershocks of the Madagascar quake are "like reverberation, like echoes in a canyon, but here the canyon is the whole interior of the earth."

But just as the narration begins to move toward a kind of soundscape of the earth's "acoustics"[6]—a resonant space of foreground and background, of sonic texture suspended in immersive depth—the earthquake sounds are brought back abruptly to their transductive and medial status as a record. Benioff says, "Now we're going to try to reproduce on this record the big San Clemente shock at nearly the original speed. It won't be audible, but by bending over the phonograph and watching the arm very closely, you will see it move from side to side, actuated by the original movement of the indicator of the seismometer in Pasadena." There follows ninety seconds of relative silence during which we hear the occasional hiss and pop of the vinyl and perhaps the faintest hum of some impossibly low-frequency tinnitus—a sound that may be buzzing in our own ears more than it is coming out of the speakers.[7] Benioff transforms the seismogram into a spectacle of pure mediality, suspending the audio output of the record player in favor of a silent, visual display of audio media as mechanical motion. The movement of the record player's arm literally reproduces the same movements of the seismograph's arm during the San Clemente quake, and if we connect the squiggles from one stylus to the next, it is as if the record needle is tracing directly the vibrations of the earthquake on vinyl. This is a record of actual earthquakes, indeed, and the album's liner notes joke that a sufficiently loud playback could result in actual earthquake-like damage to the listener's stereo equipment.[8]

The "silent" recording of the earthquake creates two opposite, yet maximally transductive moments. On the one hand, we are made to perceive—inaudibly, yet operationally—the voltage differences, mechanics, materials, and instrumental design that underlie our listening. The sonic environment shrinks down to the size and shape of the turntable, the equipmental infrastructure of our listening. On the other hand, the listening imagination expands outward to a geologic scale, transducing seismic vibrations to our listening ears that we experience no less vividly for their inaudibility. This geosonic transduction recalls Friedrich Kittler's media-theory-famous discussion of Rainer Maria Rilke's essay "Primal Sound" (1919) in *Gramophone, Film, Typewriter* ([1986] 1999). In Rilke's essay, the poet recalls a kind of science project from his school days in which his teacher and classmates construct a crude gramophone out of a bristle from a clothing brush joined to a paper cone and suspended over a manually rotated wax cylinder. The adult Rilke speculates as to what would happen if that needle were displaced from the wax grooves made from speaking into the cone and were made to trace and "playback," as it were, other grooves that were never recorded in the first place.[9] Benioff's record foregrounds the transduction of Rilke's speculative procedure, inviting us to imagine the stylus of our record player tracing the groove of an earthquake, while showing us that same stylus moving right before our very eyes, producing no sound, yet filling an imagined space the size of the planet. It is probably no coincidence that Kittler was also a fan of earthquake audification because of the extreme mixing of metaphors, materialities, and temporal scales involved in such transductive listening.[10]

In Benioff's analogy between the arm of the record player and the seismogram, the audio transduction of earthquakes brings our attention to the listening instruments themselves. Here the concept of instrument could be expanded beyond scientific sensing and recording playback to include musical instruments as well. For all the percussive hits involved in earthquake sounds, it is strings and piano keys that invite the most apt analogical and metaphorical couplings of earthquakes to earthquake instruments. The musicality, not just sonicity, of earthquakes implies that all seismological instruments—and even the earth itself—are at least potentially musical. Composer Gordon Mumma, one of the key figures in American avant-garde and electronic music, used the piano (and perhaps the earth) as a kind of geosonic musical instrument in a series of remarkable compositions based on seismographic data called the *Mographs* written in the early 1960s. During this time, Mumma worked as a research assistant at Willow Run Laboratories in Ypsilanti, Michigan, nearby to Ann Arbor where Mumma had studied at the University of Michigan. The lab, a former B-24 bomber factory, was a site for Cold War-era government research, and housed the Acoustics and Seismology Lab where two seismologists, Gordon Frantti and Leo Leverault were expanding Speeth's work on earthquake audification and bomb/quake discrimination.[11] Mumma developed an informal working relationship with the Lab, where he was able to listen to audio seismograms and was struck by the musicality of earthquake sounds.

Of particular interest to him was the temporal delay between two main types of waves caused by earthquakes: so-called primary waves (P-waves) which are compression waves that propagate through rock; and secondary waves (S-waves), shear waves that travel along the surface of the earth at slower rate, arriving to the seismometer after the P-waves with which they co-originate. Audio seismograms thus have a kind of double attack, or a staggered beginning (almost surely what Hayward had in mind with his image of a heavy door, with a latch, closing). For Mumma that interval opened up compositional possibilities: "The structural relationships of the time-travel patterns of P-waves and S-waves, and their sound reflections, had for me the compositional characteristics of musical sound spaces" (Mumma 2007: 18–19).

If we turn to Mumma's *Very Small Size Mograph 1962*, we might clarify how exactly the composer hears musical sound spaces in a seismogram. This piece is written for four pianos/four hands and was recorded by Mumma and David Tudor, the avant-garde pianist and composer known especially for his collaboration with John Cage. The recording is under thirty seconds in length and consists of single, three-note chord, but the elaborate set-up is worth noting in detail. On a Midwest tour with the Merce Cunningham Dance Company, Tudor and Mumma came across a large recital hall next to one of the dance venues. Mumma recounts:

> The recital hall had four grand pianos and Tudor suggested we do some more recording experiments with them. Those "experiments" involved elaborate spacing of both the pianos and the microphones. We made there several recordings of the very simple—only one three-note chord—*Very Small Size Mograph 1962*. The four pianos were not all well in tune, thus encouraging us to arrange them far apart. Each of us would have two pianos that we could each perform, with wide arm-stretches, at nearly the same time. We assigned ourselves the task of playing the single chord on one hand for each of the four pianos, almost, but not exactly simultaneously. For this situation we arranged ourselves so that we couldn't see each other. The performance assignment was that one of us would play the chord, and the other would respond to the sound with that same chord, then the first would respond again, and the second to the first. All very quickly so that the chord sounds became as close together as possible without being exactly together.
>
> (22)

One version of the piece consists of the four, near-simultaneous bursts of that "simple" three-note chord (B ♭, D ♭, and E ♭ above middle C[12]). The first two chords are staccato, the last two sustained for over twenty seconds until the sound dies out on its own. We hear each pianist in a different stereo channel, thus reproducing their spatial distribution in the recital/recording hall, and there is some warbling interference produced by the repeated striking of the same chord on four slightly out-of-tune pianos. The *Very Small Size Mograph* embodies seismographic data in the way it foregrounds the reaction-time of the performers, who, working out of sight of each other, have their ears and their widely stretched arms trained to respond to each other as quickly as possible, but with an inevitable delay. The concept of the piece (its "performance assignment") and the spatialization of the recording situation and the stereo playback are all structured to reproduce the lag between P- and S-waves. As intensely as the piece concentrates on a dense cluster of sound, it also aims to emphasize the otherwise barely perceptible delays and doublings within that seemingly unitary event.

The performance also reinvents the piano's function as a musical instrument into a dual functionality as a seismic sensing instrument (seismograph) and transmission medium (earth). The striking of the chord on the first piano sets up a kind of relay, with each piano then registering and responding to the initial event in real, embodied time. The pianos both transmit vibrations, as earth and rock do during an earthquake, and register those vibrations like seismographs, scrawling their readouts onto the magnetic tape of the recording equipment, as opposed to the paper printout of a Benioff seismograph.[13] In this way we could think of the *Very Small Size Mograph* as re-instrumentalizing and enearthing the piano. The performers themselves sit seismometer-like, waiting to be set into motion by distant vibrations, which arrive twice (first as a staccato chord, then as a sustained chord), even though they both originate from the same event (here, Mumma's score, or Mumma's and Tudor's "performance assignment"). The recital/recording space is transformed into something that it already is—namely a space to listen to or to record vibrations made by musical instruments—but it becomes such a space (again) only through the geosonic scale that aligns the vibratory events of the recording process with the earthy infrastructure of quakes and seismology.

The *Very Small Size Mograph* simulates not only an earthquake, but also the seismographic recording of an earthquake. Moreover, each of the elements constituting the performance (composer, performer, piano, microphone, recorder and the hall itself) function, within the transductive geosonics of the piece, like earth (as transmission medium), earthquake (as vibratory attack and propagation) and seismograph (as sensing/recording instrument), and may even take on multiple functions, simultaneously or consecutively, depending on the listener's perspective at any given point. The listener, with his vibrating eardrums and embodied audition, participates in this geosonic-tympanic equalization, too. This is *another* recording of actual earthquakes, indeed, in which "recording" encompasses the geosonic entirety of the concept, performance and recording processes. In the *Very Small Size Mograph*, the enaction of seismographic material coincides with its own infrastructure (geological, technological, embodied) precisely in the gaps between P- and S-waves, which open up "musical sound spaces" for geosonic and tympanic resonance.

Mumma's opus designation *Mograph* is obviously a play on "seismograph," and the factor of time-compression is crucial in determining the "size" (i.e., "seis-") of each *Mograph*: in addition to *Very Small Size Mograph 1962* and *Very Small Size Mograph 1963* (the former about twenty seven seconds long, the latter just fifteen), Mumma also composed *Small Size Mograph 1964* (two minutes and nine seconds); the *Medium Size Mograph 1964* (just over six minutes); and the *Large Size Mograph 1962* (just over eight minutes, the only *Mograph* written for a single performer on solo piano).[14] The larger the "size" of each *Mograph*, the smaller the factor of time-compression becomes in Mumma's compositional procedure. Less compression opens up more complex "musical sound space" in Mumma's interpretation of the seismographic data, increasing the texture and nuance of each piece in proportion to its performance time, or "size." The smaller, highly compressed *Mographs* usually have a single, quite loud dynamic range and consist of clusters of notes (single notes, sometimes chords, often staccato chords resolving to a single sustained note). Sounds are usually paired in rapid succession, recreating the dual attack of P- and S-waves at various pitch intervals, yet otherwise Mumma is careful to avoid repeated rhythmic figures, or any other structuring elements that would detract from the overall arc of each piece, which, regardless of size, traces the slow decay of a complex onslaught of sounds.

On the selection of musical pitches in the *Mographs*, Mumma writes, "One compositional decision was limiting the pitch-vocabulary and intervals for each work, allowing more attention to the complex time and rhythmic activity of the sound events" (2008: 18–19). Mumma is not attempting a systematic sonification of a particular seismographic data-set, that is, he is not mapping a range of seismic frequencies onto a corresponding frequency range of piano notes as his composition. The pitch selection is designed to convey the complexity of rhythmic events. The *Mographs* thus invert the convention of European concert music whereby rhythm is used as a means of organizing melodic or thematic material (i.e., musical pitches). By contrast, in the *Mographs* the musical notes function more like musically differentiating placeholders that allow us to perceive the rhythmed unfolding of "sound events." The *Mographs* thus generally avoid tonality and repetition so that nothing like a theme, melody or harmony can emerge. At the same time they avoid ambience and soundscapeyness. It would be unfair to call the *Mographs* an Eno-esque "Music for Earthquakes," composed with the intention or effect of smoothing out the turbulence of an earthquake into an imaginary sonic ambience. For all the meandering of the notes, intervals and chords, the rhythms feel precise and, although they are sporadic, they punctuate our attention.

The constantly shifting "pitch-vocabulary" moves the pieces forward, without cyclicality or structure, so that the overall effect is of inhabiting a dwindling texture of sonic time, listening as the sounds inevitably grow quieter, shorter and spaced further and further apart. While audio seismograms consist of a carefully processed display of earth-scale vibrations (precisely calculated via time-compression, playback speed, etc.), the *Mographs* render scale or "size" more loosely as an aesthetic experience. In a sense all we hear is *scale* when we listen to the *Mographs* arranged from small to large. The variations in the notes matter less than a sense of density unfurling into spaciousness. We hear the earth, or the earth resounds, in the duration of the *Mographs*, where the quaking of piano strings in a room merges with seismic resonance on a geosonic scale: not background music, but transductive underground music.

Part III. Earthquakes: Sounds to Music

Toward the turn of the twentieth century, as seismographic data becomes digitized and more readily processed with audio digital tools, there is a general tendency to reverse the sonic/seismic trajectory traced by Benioff and Mumma. That scientist and composer proceed from the epistemological starting point of the earthquake's musicality—its timbral texture housed within an acoustic wave-like structure—and, in different ways, use recording tools and musical instruments to break up that musicality by introducing durations of silence, uncertainty or indeterminacy. I have argued that it is in those cracks or gaps where a kind of geosonic transduction takes place, that is, where the imagination couples with techniques and tools of listening and with audio material derived from the geophysical data to listen across media, bodies and the earth. Geologic time and space are ambiguously figured and felt in these intervals between and below audified earth sounds, which aestheticize the inaudibility of the sonic material both technically and musically.

These audio seismographs use music to make the earth sensible as sound, or sonic events. Neither audio seismograms nor *Mographs* constitute soundscapes: they are discrete, particular geosonic events transduced by machines, bodies and the listening imagination. They mark a moment in time as a juncture, otherwise unimaginable or imperceptible, of human and geologic scales of time and space. They also trace, often literally, the fault lines where human technocultures encounter and reimagine their geologic, material substrate. Yet moving into the twenty-first century, several artists and musicians are more inclined to use seismic data as a more readily exploitable musical resource. Digitization—with its capacity to work with vast data sets of synchronized global seismometric networks, and to stream data in real-time and play it back automatically via custom algorithms—replaces the tape-based transductive processes that previously converted earthquake "music" into singular sounds. The easier access to geophysical data and the more or less instantaneous conversion of data into sound lend themselves to seismic soundscaping, a kind of re-backgrounding of the underground. I turn now to some key contemporary digital seismic installations and performances that offer a kind of tectonic shift in the geosonic imaginary, that is, a dubious digital environmentalism of earth as immersive background, ambience and atmosphere.

Ryan McGee and David Roger's work on the "Musification of Seismic Data" (2016) highlights new digital techniques and platforms for earthquake audification that the authors have put to use in contemporary musical compositions and media installations based on seismographic data. Beyond techniques of time-compression and frequency doubling associated with tape manipulation, McGee and Rogers use granular synthesis and phase vocoding to alter the duration, pitch, and timbre of the percussive hits of "natural" earthquake sounds. Granular synthesis (which I discuss in the context of 24/7 audio culture in Chapter 3) involves fragmenting digital sound samples into smaller pieces or "grains." In the case of synchronous granular synthesis, the grains are essentially copied out a number of times and pasted back together to increase the duration of a sound without altering its pitch. The technique has the effect of "repeatedly emphasizing each grain, which emphasizes the unique decay of each earthquake" (McGee and Rogers 2016: 2). Asynchronous granular synthesis is a similar process, but which imbricates different grains, rather than repeating and pasting together the same grain, with the effect of "creating stuttering rhythms unique for each event" (2). The spaciousness and staggered attacks of Mumma's *Mographs* here become, respectively, a kind of digitally smoothed time-stretch and a regularized rhythmic pulse. Phase vocoding is another technique that breaks apart, filters and resynthesizes digital sound samples. It can be used to edit out undesired frequencies to "de-noise" a sound: "This de-noising can be taken to extremes to leave only a few partials in each sound, ultimately producing unique tones and chords for seismic events" (2). Vocoding can thus introduce clean tonality into a noisy frequency spectrum so that we can hear precise musical intervals (minor thirds, fifths etc.) emerge out of the audio spectrum.[15]

McGee has produced two electronic compositions, "Haiti" and "Christchurch" (2012), each based on seismographic data from the deadly, high-magnitude quakes in those cities in 2010 and 2011, respectively. The opportunities for stretching, layering, and filtering the sounds create a distinctly musical ambience characterized by an interaction of "instrumental" timbres and harmonic intervals, as well as moments of pulsed rhythm that beat a kind of 4/4 time. One user on Soundcloud (where McGee has uploaded both recordings) writes of the "Haiti" piece, "It's like a surreal dubstep sampler;" and another user writes, "that drop is filthy," referring (approvingly?) to a dubstep-like beat drop in the "Christchurch" piece. If Benioff was trying to make the seismographic data sound "like" an earthquake (whatever that might mean, and even if that meant recourse to an inaudible sound), then McGee and Rogers clearly want to make earthquakes sound not just like music, but like a specific subgenre of contemporary electronic dance music.

"Haiti" and "Christchurch" played continuously as part of a seven-month installation titled "DOMUS" at Materials and Applications, Los Angeles (October 2014–April 2015). The authors describe the project as "an experimental architecture installation incorporating spatialized seismic sound and light within a [two-storey] hexayurt [equipped with] 6 mid-range speakers, 4 subwoofers and a 360 degree LED pixel chandelier" (McGee and Rogers 2016: 2–3). The loud, low-end playback resonating within the space of the yurt emphasizes what they call the "seismic nature" of the compositions in the context of a "biophilic design model that responds [*sic*] and reflects the natural world" (3). Like Benioff's *Out of This World*, the installation aestheticizes inaudible data. But where the reality effect of Benioff's audifications derives from explicit and implicit ideas about the materiality of sound and recording, the reality effect of the DOMUS installation appeals instead to "nature" and to unscrutinized sonic expectations of earthquakiness. DOMUS simulates what people think an earthquake sounds like, rather than using sound to make people think about earth(quakes), full stop. The assumption that audio seismicity has a "nature" or is "natural" and would thus slot into our culturally constructed listening habits of what "nature" sounds like is as problematic as the aesthetic possibility that seismicity also happens to sound like dubstep. But in effect both propositions downplay, if not outright ignore, the transductive nature of audification. They emphasize either a smooth, regularized sonic product (dubstep) or a soundscape of "natural" earthquake sounds, rather than foreground the indeterminacy of geosonic processes that cut across earth, instrument, body, and imagination, simultaneously making, equalizing, and unsettling all the elements in the transductive loop.

McGee and Rogers have also developed an ambitious project called "Sounds of Seismic" (SOS) which they characterize as an "infinite computational earth system soundscape" (2016: 4). SOS uses custom software applications to collect data streaming in real-time from global seismographic monitoring networks and automatically generate live-streaming music out of the data, with the help of algorithms processing the data through the synthesis and vocoding techniques outlined above. Seismic data remains "the original, sole sound generator" for this automated, 24/7 data-processed soundscape. While this may sound like the ideal seismic audification of the twenty-first century, the conceptualization of the project brings us oddly back to the mid-twentieth century avant-garde. McGee and Rogers cite Pierre Schaeffer's tape experiments and *musique concrète* as inspiration, and more significantly, John Cage's *Variations VII* ([1966] 2008), an ambitious performance/installation staged at the Armory in New York and incorporating live electronic transmissions of deep-space radio waves, ionospheric "whistlers" and Bell telephone lines, mixed with on-stage amplifications of the sounds of blenders, fans and footsteps on the contact-microphoned stage platform, not to mention amplified readouts of electroencephalographs and cardiographs of some of the performers (which included Cage himself and David Tudor).[16]

The Cage piece is a multi-scalar festival of intra-corporeal, infrastructural, everyday, and extra-planetary sounds premised on the idea that using technology to tune into otherwise inaccessible sounds is of inherent aesthetic value.[17] Rather than question how we perceive these sounds and what kinds of technocultural and ecological relations they enact, the piece works like a conventional concert hall performance: virtuosically choreographed sounds produced live by trained technicians working with specialized instruments resonate in a large interior space before a paying audience. Only here the musical raw material to be appropriated is "natural" (if highly mediated), rather than "musical" in any conventional sense. The piece takes for granted a vast background or atmosphere of hidden sounds that are presumably available, first as electronic signals, then through a transparently understood process of transduction, as (musical) sounds.

Such an atmosphere of availability is problematic on its own terms (which does not, I should stress, necessarily diminish the aesthetic impact of Cage's piece), yet becomes even more problematic in case of SOS, when it goes underground, that is, when seismic sounds are literally and metaphorically grasped and instrumentalized as musical atmosphere. McGee and Rogers conclude the presentation of the work on musification with a nod toward the ultimate background, instrumental music (which I discuss in Chapter 5): Muzak, writing "The ultimate goal of this research is to create a generative, dynamic audification-based musak [*sic*] which can highlight resilience and awareness of the natural world in which we inhabit [*sic*]" (4). Given that historically Muzak is linked to environments of docility and distraction rather than resilience and awareness, this aesthetic choice is as contradictory as the idea that we can be brought into contact with the earth by an "infinite computational system."

A project like the SOS is marked by a number of contradictions: it draws inspiration from the mid-century Cagean assumption that because certain distant signals can be captured as sounds, therefore they also *should* be. It also wants to smooth out and regularize those jagged signals into music, while preserving their "natural" qualities at the same time. To return to Connor's argument from above, music's claim-staking to atmosphere and ambience overlooks/overhears the "down to earth" trafficking between foreground and background (i.e., transduction) that any conceptually or experimentally minded music must continually negotiate. Just as Eno suspends the very sonic environment he composes ambient music for, so does a project like SOS bracket out the earth—and the earthly infrastructure of listening—by transforming it into a database of Muzak-able material. In SOS and similar projects, twenty-first century earthquakes are thus assimilated, via a detour back to the mid-twentieth, to the contemporary technocultural model of the networked computer: they are streaming in real-time, easily and ubiquitously accessible, and they increase our awareness of "nature" insofar as they serve as an ambient soundscape available 24/7.[18]

One notable exception is Steve Roden's intriguing sound-installation *ear(th)* (2004), in which the artist uses digital tools to transform an interferogram (a contour map of the effects of an earthquake on the earth's surface) into a looping piece of music. Roden constructed a tunnel-like wooden form (8′ tall, 16′ wide, 24′ long) and installed 80 robotic arms that strike 80 glockenspiel bars on top of the structure, with each individual arm programmed to respond to one of 80 data points from the interferogram. The wooden form is like a giant ear, with its own tympanic structure (robot arms tympanically hammering on glockenspiels); and its build specifications echo the musical octave, with the number 8 being the base proportion of the wooden structure (8/16/24) and the sounds it makes (8 notes, 80 glockenspiel bars, and 80 data points). The ear metaphor, the arbitrariness of the Western musical octave, and the bulkiness of the built structure in the gallery space (suffused with contrastingly delicate glockenspiel tones) all interpose a conspicuous technical setup between the listener and the sounds of the earth. Rather than exploiting a putative essentialism between ambient listening and earthquake sounds, Roden has a humbler, yet more expansive aim in mind: to create "a controlled listening situation" based on "a kind of alchemy that allows the material to generate something completely connected to, yet seemingly distant on the surface from, its source." Sonification, for Roden, means foregrounding transduction, inviting a reflection on geosonic mediation precisely because the "[geophysical] information is determining something completely different from its original intentions" (Roden 2004).

Another artist, Moon Ribas further adapts the earthquake to another contemporary technocultural model, that of the online mobile device. Ribas, a Barcelona-born collaborator with Neil Harbisson (discussed in Chapter 4) and co-founder of the Cyborg Foundation, is a self-proclaimed cyborg artist who works with seismographic data transduced through a sensory prosthetic-cum-implant. Ribas initially used a wearable wrist sensor that transmitted seismographic data to her body via vibrations in the wristband, creating what she calls a "seismic sense." But as she mentions in a 2015 TED talk in Munich, "I didn't want to wear my sense, I wanted to have my sense." Since then, a version of the sensor has been implanted directly into her left arm, just above the elbow, linked to the internet via Bluetooth. A second implant in her right arm transmits vibrations from seismographic monitoring systems on the moon, so that in addition to the "earthbeat" (Ribas' term for the heartbeat-like buzzing that represents the real-time aggregate of global seismicity linked to her implant), Ribas can also sense the fainter pulsations of moonquakes. In a 2016 TED talk at McGill University, Ribas called herself a "senstronaut" whose lunar- and geoseismic senses allow her "to be here and in space at the same time." Like

Harbisson, Ribas promotes her cyborg identity through the aforementioned TED talks and other YouTube-friendly presentations/interviews. Her signature performance piece is titled "Waiting for Earthquakes" and involves spontaneous choreography—Ribas trained as a dancer—in response to vibrations of the seismic sensor as transmitted in real time during the performance time. In a highly edited version of a performance of the piece (uploaded to YouTube in 2013), we see Ribas dancing solo with apparently no musical or sound accompaniment. She wears a sleeveless dress to enhance the visibility of her pre-implant wrist sensors, a visibility emphasized throughout the performance with vigorous, flailing arm movements. Early on in the video, we see Ribas falling to her knees, torso undulating smoothly, or sometimes jerking spastically, as she attempts to jump up, only to fall back to her knees. There is a vague sense of narrative in the performer's struggle to stand, to locate some kind of choreographic *terra firma*. The piece overall is characterized by fluid movements and abrupt transitions, occasionally pulsed, yet without a persistent rhythm or pattern.

In a 2017 interview, Ribas explains, "My artwork is happening inside my body. I'm the only one who perceives it. So in order to share my experience, I create external artwork related to my new sense" (Ribas 2017). Related to "Waiting for Earthquakes," Ribas has also created performances of "seismic percussion, where the score and the rhythm I play [are] dictated by the rhythm of the tectonic plates. In this case, earth is the composer of the piece" (Ribas 2016). As with Harbisson, the setup of the cyborg performance involves conjoining the extreme solipsism of an absolutely interior, subjective experience with the conceit of the unmediated operation of "technology" or, in Ribas' case, "the earth." In "Waiting for Earthquakes," we see, theoretically, the earth's seismicity transduced in real time into the movements of the dancer's body. In the percussion piece, we can hear the cybernetically enhanced rhythm of the "earthbeat" communicated to our eardrums by Ribas, as interpreter and conductor of the earth.

Setting aside the looming technical questions and doubts that such performances might bring up,[19] we encounter in Ribas' work a version of earthquake audification that wants to bypass the ear altogether, vibrating and resonating directly on/in the body. Her prosthetic displaces the tympanism of the eardrum onto a tactile interface. Earthquake sounds here become a prototype of a wearable device or a mobile app. Even if we take Ribas for her word and believe that a small, magnetic vibration in her arm is capable of conveying, in real time, the massive, and massively complex data sets of global seismometry (which involve thousands of sensors, each of which move vertically, north–south and east–west, and register vibrations from multiple, simultaneous events, time-traveling via reflection and refraction through a deep planetary structure that geoscientists still do not fully understand), we still have to accept the proposition that earthquake sounds are not only streaming 24/7, but that they should also be integrated into our bodily senses and rhythms as well, like a Fitbit or an Apple Watch. The overall effect is, as in McGee and Rogers' work, a superficially "seismic" environmentalism: "I feel closer to nature because I can feel my planet. And I feel closer to other animals that can perceive earthquakes, too. Adding new senses to the body can help us to rediscover the planet where we live [and] bring us closer to the planet, to nature and to other animal species" (Ribas 2015). And from the McGill talk, we hear the slogan, "Let's try not to change our environment and be brave enough to change ourselves" (2016). Like Harbisson, Ribas advocates a cyborg identity in which the individual aesthetic imperative toward bodily enhancement is imagined as a step toward ecological awareness and a kind of fight against climate change. If we all just had the bravery to change our bodies, we could both preserve the environment *and* feel closer to nature, other species, and the planet all at the same time.

What these recent experiments in digital seismicity suggest is a pseudo-environmentalist updating of the mid-century aesthetics of audio seismograms. At the technical level there is a kind of paradigm shift from the tape-based tympanism of Benioff and Speeth's recordings and Mumma's pianistic reimaginings to a computer/mobile device paradigm. Sounds, seismic or otherwise, no longer have to be recorded (by a machine, or an ear) in order to exist and be listened to. Rather, sound is reconfigured as a matter of algorithmically facilitated access or connectivity: seismic sounds exist to the extent that, at least potentially, they can be streamed in real-time 24/7, as earth-generated Muzak or as Stelarc-esque prosthetic ears vibrating on our arms. But such sounds are literally out of their element in that they presume earth can be listened to as atmosphere, or that the human body can "beat" like a planet. Transduction is bypassed conceptually (but not technically) in favor of a rhetoric of environmental awareness and connectedness to nature and the earth. Where Benioff and Mumma emphasized attack, decay and inaudibility as fundamental to the idea of seismic playback, digital seismicity (with the exception of Roden's more indeterminate processing in *ear[th]*) connects to a smooth flow of continuous access. Perhaps the imminent threat of nuclear destruction—and indeed, underground nuclear explosions were being audified along with earthquakes—marked those Cold War-era sounds with a sense of inevitable finitude; or perhaps the finitude of magnetic tape itself contributed to a mid-century geosonic imagination of earthquakes as singular, time-stamped events. And perhaps the diffuse, ongoing phasing of climate crisis in the twenty-first century lends itself to a reimagining of earthquakes not as dramatic individual events, but a vast, global soundscape without beginning or end. The difficulty with the latter proposition would be its denial of the literally earth-devastating impacts that underlie global 24/7 digital culture, its recasting of global seismicity as a soundtrack for climate crisis.

The point of my intervention is not necessarily to condemn or praise one historical moment over another, but rather to trace this fault line in the geosonic imaginary, where one set of technical, scientific, aesthetic, and geopolitical circumstances collides with another. My conclusion is theoretical, more than critical: there is no sound of the earth per se, only the myriad and fleeting configurations of technology and imagination by which we transduce a necessarily partial and fleeting earth. I do not mean to say that the earth does not exist, that there is no outside to the human technocultural habitation of the planet, and from there to make a wholesale critique of "earth" as inflated anthropomorphism. But I do mean to say that any facile geomorphism—any claim to authenticity, organicity, or priority resulting from a technological or aesthetic set-up on/in/under/as "the earth" does not, indeed cannot reveal the earth, but only refers us back to the aesthetic and technological crafting of an environment in historically specific, contingent ways.

3

A Planet Made of Beethoven

Audio Stretching, Transductive Listening, and 24/7 Aesthetics

Are sounds just sounds or are they Beethoven?

John Cage (2011: 41)

Whenever we are immersed in a soundscape, the actual earth disappears, but, as I argue throughout *Geosonics*, it can be partially recovered through transductive listening, that is, through critical attention to the sonic processes that mediate between bodies, instruments, and the earth-as-interface. At the same time, transductive listening takes place within historically and technoculturally contingent moments (i.e., geosonicks) that always interface with a *particular*, version of the earth, never the earth as such. In the previous chapter, I discussed the geosonics of earthquakes and identified a kind of paradigm shift in audio seismology from mid-twentieth century, tape-based audification techniques to twenty-first-century digital synthesis. The shift is technical, moving from the finitude of working with tape to the fluidity of data aggregation and simulative software tools, but also aesthetic insofar as different instruments express the particular aesthetic and epistemological premises built into their material affordances. In the case of digital audio tools and networks of real-time seismic monitoring, we listen to an earth that has already been datafied and can thus be displayed as digital sound and music seemingly without medial translation. The digital seismic earth appears already as a 24/7 soundscape via the tools scientists use to listen to it.

Moving into this chapter and anticipating the discussion below of Jonathan Crary's *24/7: Late Capitalism and the Ends of Sleep* (2013), I explore connections between the earth and the global economic system of digital capitalism, positing the soundscape as an intermediary between the two. 24/7 society forces the temporal horizon of everyday life (the 24-hour day) to align with the scale of contemporary digital capitalism (already operating globally around the clock). Even though, as Crary notes, people cannot work or shop 24 hours a day, they are nevertheless coerced, as neoliberal subjects, to live their daily lives as if they could, and thus to live virtually according to a human-incommensurable scale. 24/7-ness becomes a specifically geosonic problem when sonic media step in to smooth out that underlying incommensurability, aestheticizing the gap between an earth-scale system (global capitalism) and individual daily experience into a 24-hour sonic environment or soundscape. As with shopping and working, we can't listen attentively 24/7, yet contemporary listeners are coerced to do so theoretically via the market of products and services that facilitate the possibility of 24/7 listening-through-online-connectivity.

I attempt to tune into and listen through the 24/7 soundscape by focusing on an artwork called *9 Beet Stretch* (2002) by Norwegian artist Leif Inge. The work uses the digital process of granular synthesis (mentioned in Chapter 2) to extend and smooth out the duration of a recording of Ludwig van Beethoven's Symphony No. 9 to a length of twenty-four hours. There is no intrinsic formal reason why the audio stretching of the Symphony No. 9 should be twenty-four hours long. The otherwise arbitrary choice of duration attempts to fuse two kinds of immersion: immersion in the stretched soundscape of the symphony with immersion in/on a planet that rotates once every twenty our hours. Geosonically, the immersiveness of the listening experience entails the temporal earth-scaling of the listener, remediating the concert-hall aesthetics and acoustics of the symphony on a planetary scale. At the same time, the immersiveness of that experience displaces or omits the infrastructure of audio streaming and 24/7-ness, namely digital capitalism and the geopolitics of labor. I attempt to draw into critical focus the moments where the immersive listening experience reveals/conceals its economic underpinnings.

Inge's installation suspends the listener in a geosonic elsewhere that is uneasily situated between the transcendent musical spaces of European art music and the grim reality of 24/7 environments sketched out by Crary. The *Stretch* pulls us impossibly close to the materiality of the orchestra and the voice; the archival reality of a recording; the imaginary mechanics of a digital process; and the compositional structure of Beethoven's masterwork. At the same time the piece constantly pushes us away toward a soundscape of "sounds in themselves" that we may experience variously as sublime, ambient, annoying, boring, et cetera. In search of aesthetic solutions at formulating the seeming incommensurability between a technical process of nonstop duration and the human act of listening, I build on John Cage's notion of indeterminacy ([1966] 2008; 2011) and Mark B.

N. Hansen's (2004) theory of affective embodiment, critically situating the latter against the 24/7 world of digital apparatuses outlined by Crary. I conclude, somewhat anachronistically, with a discussion of Adorno's reflections on Beethoven and musical time from the late 1940s. Although Adorno does not employ the term soundscape, he is a thinker keenly aware of music's aesthetic power to reflect and refract social forms of domination. As such, his insights into sonic time as a modality of capitalist domination presciently listen ahead to the critique of the 24/7 soundscape that I offer here.

Nine minutes into the final choral movement of Beethoven's Symphony No. 9, we encounter a remarkable rift in the musical fabric. It consists first of fermata—in which the orchestra and chorus sustain a single chord for an extended interval—followed by an equally long silence, or caesura. In the moments leading up to this break, we hear the delicate elaboration of that famous "Ode to Joy" theme by the solo singers, interspersed with grand entrances of the entire choir. The orchestral playing emphasizes the upbeats of each measure, moving against the dominant flow of the vocal theme, and creating a powerful, if also powerfully imbalanced sense of dynamism and energy.

Just when it finally seems that the orchestra and the voices will coincide on the downbeat and restore balance to the piece, there are two abrupt key changes. Repeating the lines "und der Cherub steht vor Gott," the singers resolve a cadence in the tonic key of D, but swiftly move back to the dominant A and then, unexpectedly, to F, where the fermata takes hold. In the first shift to A, it is as if the piece attains a new height, and though we may not have expected it, the move from tonic to dominant works harmonically, and the elevation of the overall tone works musically as well. Yet the shift from A to F is more radically unsettling. All the upper voices hold the same high note, the A, but the lower registers all drop to F. It is as if the bottom has dropped out from under the piece, as if the newly modulated A had become distorted, stretched itself to a kind of breaking point. And it is precisely here that the piece rests, sustaining that F chord for seconds that may feel like hours, forcing us to question if this is the new musical ground on which the piece will stand, or if it is a mere interruption or deviation which will eventually take us back to familiar territory. After the challenging pause comes an equally challenging silence, refusing to resolve any of our questions, suspending us in a stunning cessation of musical time.

If we add to this pause not just the preceding nine minutes of the choral movement, but also the entire forty-odd minutes of the first three movements of the piece (already in themselves longer than most entire symphonic works), we begin to feel the truly crushing weight of musical energy and time that has come to rest on this moment of sustain and pause. An entire piece has derailed itself, and the listener—whether novice or expert—may, at each listening, have virtually no idea how the piece could continue after such a monumental disruption.

I have just written nearly 500 words about a passage that occupies 12 bars in the score and roughly 30 seconds of performance time (Beethoven 1997: 126–7). It is one of the great privileges in writing about music that you can allow yourself an expanded time for reflection and analysis (not to mention all the literary resources of metaphor) to express thoughts on mere seconds of actual music. Writing about music depends on the fantasy that you can fully separate the aesthetic time of the piece in question from the empirical time of its performance, and dwell in a frozen moment of musical time, safe from the ticking-away of the clock that only moves you further and further away from the sounds that initially captured your attention.

Listening to music is far more difficult than writing about it because, from the position of the listener, the piece always rushes ahead in empirical time, and indeed, in musical time as well. If we choose to linger in our minds on a particular passage, we risk missing what follows it and thus lose the flow of the piece. And indeed, the more I think about this particular passage without regard for the whole, the further I am from what it actually sounded like, the more I am filling it out in my memory. Even if I use a recording to find it and play it back, I would notice that breaking and resetting the flow of empirical time also compromise the aesthetic effect of the musical time. When I play back a passage half a dozen times, it loses the effect it has when it comes only once in the temporal flow of all that precedes and follows it.

This deficiency in listening to music, its built-in belatedness, is of course also what makes listening and re-listening to music endlessly fascinating and productive for thinking. The belatedness is, as I argue in the Introduction and Chapter 1, also intrinsic to the "always again" remediation of sound through listening. And, to come to the technique of audio stretching, and the *9 Beet Stretch* in particular, it also leads to a series of questions: What if we had more time for listening to music while we were in fact already listening to music? What if we could adjust the ratio of aesthetic time to empirical time in a piece of music so that we could more substantively integrate the time of our musical reflection into the actual empirical time of listening? Rather than speeding up our thoughts, what if we could, in effect, slow down the music, but in such a way that we could still hear it and think with it at the same time? And what would such a piece sound like anyway?

Norwegian artist Leif Inge's *9 Beet Stretch* is a composition/installation that digitally "stretches" the length of a recording of Beethoven's Symphony No. 9 to a duration of twenty-four hours. Conceived in 2002, the work premiered in Madison, Wisconsin, in 2004, with subsequent performances in Vienna, Shanghai, New York, and Toronto, as well as a 24/7 webcast streaming since 2005 and a free app available for download. The "stretch" uses a technical process called granular synthesis to essentially break apart, replicate and smoothly recombine the digital samples of a recording with the effect of extending duration without pitch distortion. The listener can hear Beethoven's symphony extremely "slowly," yet without losing the tonal frequencies and other sonic qualities of the recording played at its normal speed. The result is an impossible listening experience, a clash of temporalities in which a piece originally scored for a performance of roughly seventy minutes can be experienced as a twenty-hour-hour event.

The immense scale of the *9 Beet Stretch*, along with the unrelenting, "time-release" uncanniness of hearing accurate musical pitches vibrating with physically impossible slowness, force the audience to reconsider its fundamental assumptions not only about sound, time and listening, but also about the limits of live performance and the hidden depths of recorded sound. Yet, as an experimental work, Inge's piece stands out for two more reasons: (1) its insistence on a *fixed*, if radically extended, performance time and (2) its explicit reintroduction of classical European musical content into the field of experimental music and sound art. In turning away from temporal indeterminacy (as theorized and practiced by John Cage, among others) and in turning back to a nineteenth-century masterwork, the *9 Beet Stretch* seems to demand an uneasy synthesis of two incompatible sonic aesthetics: the structured, time-objective, individual-expressive qualities of Beethoven's composition with an ambient, time-arbitrary sound environment. Inge seems to ask us to rethink the fundamental nature of listening while, at the same time, listening *to* a fixed, and indeed quite famous piece of music; or, to put it in the terms I used previously, to integrate listening and reflection into a single sensory-intellectual experience that unfolds in a continuous timeline.

Yet the integration of listening and reflection already has a powerful precursor in twentieth-century musical aesthetics, namely John Cage's notion of indeterminacy, and it is Cage's playful, yet incisive distinction between "just sounds" on the one hand and "Beethoven" on the other that marks the point of departure for my notion of transductive listening. For Cage, the single proper name "Beethoven" evokes the entire tradition of European art music, with its emphasis on individual authorship, teleological form, subjective expression, and the performance conventions of the concert hall. In seeming opposition are "just sounds," that is, the notion of "sounds themselves" produced through compositional techniques involving chance and experimentation, as well as performance practices that subvert the aesthetics of the concert hall. Indeterminacy, for Cage, names the processes that sought to liberate time, space, and indeed sound itself from the constraints of Western music,[1] processes that culminated famously in *4′33″* (1952), the "silent" piece that effected a redirection of audience attention from the structure of performed music to the incidental ambience of the environment. Rather than listen merely to "Beethoven," one can listen to literally any sound, and discover a hidden musicality, an implicit aesthetic dimension to any experience that emerges out of the perceptual act itself. Indeterminacy is, then, as much a mode of listening and a form of attention as it is a process of composition and performance. When Cage composed a kind of sequel to *4′33″*, namely *0′00″ (4′33″ No. 2)* (1962), he expanded the perceptual field from the hall to the world of everyday life, inviting an open-ended, potentially infinite merging of perception, aesthetic experience, and the act of living under the auspices of indeterminacy.[2]

In his 1958 lecture "Composition as Process II: Indeterminacy," Cage writes, "It is high time to let sounds issue in time independent of a beat in order to show the necessity of time" (2011: 40), a somewhat paradoxical statement I would interpret as follows: Cage would reject from the outset the duality of aesthetic to empirical time, insisting on only a single, unified but indeterminate time (what he calls "the necessity of time") in which musical sounds should be organized, performed and heard. The more a musician or composer work in fixed units of time and under the compulsion to make sounds go somewhere in time—to endow them with a temporal teleology—the more time itself recedes from the experience of composition, performance or listening. Rather than rendering time inaudible by subordinating it to musical form, composers should strive to make time something we can hear along with sound.

Some of the more notable compositions to follow in Cage's theoretical footsteps emerged from the Fluxus moment, such as Takehisa Kosugi's composition *South No. 2 to Nam June Paik*, in which the word "south" is pronounced extremely slowly, for a duration of at least 15 minutes; or La Monte

Young's *Composition 1960 #7*, for which the only scoring is to hold a B and an F sharp "for a long time." Douglas Kahn has remarked that these Fluxus experiments reveal that "any single sound contain[s] exceedingly complex processes of production, of internal configuration: that a single sound's interaction with corporeal and environmental space transform[s] it from one moment to the next; and, therefore, that a simple musical structure of repetition or sustainment [is] not simple at all" (2011: 37). In other words, the indeterminacy of time, or merging of musical time and empirical time, lets us hear a hitherto unheard-of complexity in the production of simple or repeated sounds. Musical content organized in time in fact obstructs our sensation of sound, and a composition for which musical time is irrelevant, or which fuses musical and empirical time, gives us a wider, deeper knowledge of sound's materiality.

If Cagean indeterminacy blurs the distinction between art and life, between a world of mere noise (including "Beethoven") and a world of silence populated by aestheticized sounds, then transductive listening introduces a third term into the fluid dualism of "Beethoven" and "just sounds:" the technical process itself (here, audio stretching) which both enables our listening and stands in as its object at the same time. I would argue, in other words, that when we listen to the *9 Beet Stretch*, we listen to three things at once: Beethoven, just sounds, and audio stretching in and of itself. The stretch process paradoxically virtualizes *and* concretizes the sounds and the music into an excess that our listening ears sample repeatedly in a frustrated attempt to connect with a process that defies the psychoacoustics of listening.

Ordinarily, when we listen to the complex sounds that compose a piece like Beethoven's Symphony No. 9, the ear is not so much absorbing the totality of the sound, but rather constantly filtering, simplifying, aggregating or, to use Aden Evens' (2005: 1–7) term, *contracting* the infinitely more complex layerings of sound waves that make up the tone of a single violin, let alone a full orchestra, or the concert hall itself.[3] Listening both reduces and synthesizes a far more complex materiality comprised of a continuum of sound and noise, contracting it into the timbres that we may experience as pleasing, and the melodic or harmonic configurations that we find expressive. Yet when we listen to a stretched recording, there takes place a particular kind of reversal of the usual "contraction" that our listening ears make. Rather than contract an infinitely complex acoustic materiality into distinct timbres, that is, rather than move from noumenal noise to organized sound, audio stretching allows us, or forces us, to move from organized sound back to noise

As we listen in to a given segment from the *9 Beet Stretch*, we may hear sounds that we initially recognize as human voices, violins, trumpets, etc., but that initial recognition slowly dissolves into an indeterminate timbral texture. The individual tones that are, from the compositional point of view, the building blocks of the piece, are revealed to be unstable, uneven, and highly composite acoustic/digital entities. In an unstretched audio world, we may know this to be case—that erratic, recalcitrant, and boundless sounds compose the acoustic reality underlying the fiction of expressive music—but we cannot hear it directly. Yet in a stretched audio world, we can hear the very material conditions of hearing itself; we can listen to the psychoacoustics of listening via a technical process that both reverses *and* plays back the very process of listening.

When indeterminacy is accomplished in this way—not at the level of composition or performance, but rather through a technical process of remediation—we encounter the form of transductive listening characterized not just by the fluidity of music/sound, but also that of object/process. To return, in stretched form, to Beethoven's 9th, and in particular, to the fermata of that dramatic key change I described earlier, the "sudden" shift to the key of F would take on a completely different character, approximating the purely sonic and leaving behind the merely musical. We would hear not just a dozen sopranos hitting a high A; we would hear the protracted whirring and warbling of dozens of voices as they searched, in the microseconds of their vocal attack, for the precise pitch, followed by the meandering flux of volume and pitch as uncountable hidden rhythms appear and disappear in the micro-fluctuations of the voices and the ever-so-lurching crossfade from one granulated sample to the next. Likewise, instead of hearing the strings drawing double *forte* on the F chord, we would hear (or imagine ourselves hearing) the skittering of hundreds of thousands of strands of horse hair pulled at minutely different speeds with minutely different degrees of friction and tension across dozens of steel strings, each string scraping and coming to life in a sonic cluster of vibration that we are otherwise accustomed to call, merely, "playing in tune." We could similarly break down the sonic contribution of each section of the orchestra, as that extended F chord surges and trembles in prolonged transition. Even the "sudden" break from sound to silence in the ensuing caesura would be rendered into a gradual process, a turbulent gray area of sound decaying, unevenly, in fits and starts, into a silence that lasts for minutes, pulsing and hissing with echoes and room-tone. The abruptness of this passage of the symphony—its syncopated build-up, rapid key changes, and the fragmentation of the fermata—would be transformed into a murkily gradual transitional process, the stark musical contours softened into an ambient flux. The fixed, written score of Beethoven would thus yield an excess of spontaneous sounds, rhythms and textures that co-originate with the performance of the notes on the page, but could not themselves be reduced to Beethoven's composition.

But we would go too far to say that the *9 Beet Stretch* effectively transforms "Beethoven" into "just sounds" because we are not dealing with infinite or variable duration, but rather a fixed, if distorted scale of listening (in which one second of Beethoven's 9th is stretched to roughly 22 seconds of *9 Beet Stretch*). In other words, the piece still moves "forward" in time and as it does so, we hear not so much music-as-sound, but rather the becoming-sound of music. Within the fixed temporal parameters of the piece, we can hear the hidden depths, the micro-tonalities and accidental rhythms of the stretched recording, we hear the sound *within* Beethoven's music, not just the sound *of* it.[4] For this reason, it would also be unfair to assert that the *9 Beet Stretch* completely eliminates Beethoven as the author the work, for we are still immersed in his composition, and are merely being asked to listen to it on a different scale or listen for different things in it than we other otherwise would.[5]

Yet the immense scale of the *Stretch* would seem to defy the human ability to attend to the interstitial excess of sound for its programmed duration of 24 hours. Gaston Bachelard's notion of "intimate immensity" might give us a way of formulating the symmetry between the technical process of audio stretching and our process of listening. Bachelard describes the experience of immensity as a daydream, as a particular form of contemplation in which the thinker does not approach a particular object with his consciousness, but rather recedes from it and finds himself in a distant "elsewhere," the space of the daydream. He calls this state "phenomenology without phenomena; or stated less paradoxically, one that, in order to know the productive flow of images, need not wait for the phenomena of imagination to take form and become established in completed images. In other words, since immense is not an object, a phenomenology of immense would refer us directly to our imagining consciousness. In analyzing images of immensity, we would realize within ourselves the pure being of pure imagination" (1994: 184). Adapting Bachelard's language to listening and sound, we could say that Inge's piece allows us to recede from Beethoven's masterwork, situating us in a sonic "elsewhere" in which we listen to an unformed, yet constantly forming flux of sounds, and, in effect, listen to listening itself in the interstices of music and sound that Inge's piece reveals.

Mark B. N. Hansen (2004: 197–232), in his discussion of Robert Lazzarini's *skulls* (2000), elaborates a similar model of the feedback loop of perception and reflection (here mediated by a digital process). But whereas in Bachelard the redirection of perception folds back on the imagination itself in a strictly imaginative elsewhere, Hansen describes how a short-circuit in digitally mediated perception takes place paradoxically in the body, not the mind. *Skulls* is a sculptural installation consisting of four skull-like objects mounted at eye-level on four walls of a room. The objects are physically realized sculptures based on digitally distorted CAD scans of an actual human skull. They appear, as Hansen notes, initially as anamorphic images that recall Hans Holbein's famous anamorphic skull in *The Ambassadors* (1533). Yet, unlike Holbein's skull, Lazzarini's *skulls* presents *three*-dimensional objects derived from a two-dimensional distortion in the computer. As such, the very physical depth of the objects constantly interferes with the eye's effort to undo the distortion: "*Skulls* confronts us, in short, with a spatial problematic we cannot resolve: with the 'fact' of a perspectival distortion that can be realized (and corrected)—and that 'makes sense' visually—only within the weird logic and topology of the computer" (Hansen 2004: 202).

For Hansen, *skulls* "furnishes what amounts to a cipher or index of a process fundamentally heterogeneous to our constitutive perceptual ratios" (204). And yet the strangeness of this irruption of the digital into the human sensorium produces an embodied, rather than abstract response. We *feel* the skull-like forms in our bodies precisely because we cannot see them in human-scaled visual space. The work "functions by *catalyzing* an affective process of embodied form-giving, a process that creates *place* within our bodies. And since it is through such a *creation* that we get a sense for the 'weirdness' of digital topology, we might well think of it as a correlate to the impossible perceptual experience offered by the work" (203, emphasis in original).

Skulls is exemplary of the reconceptualization of the digital image that Hansen develops throughout the book, whereby the image no longer refers to fixed objects or forms, but rather to the very perceptual process by which the body "gives form to, or *in-forms* information" (10). Paradoxically, it is precisely as a digital *image* that *skulls* is also essentially a "post-visual" work because the response it provokes is proprioceptive, haptic, and bodily rather than the disembodied abstraction that one might initially associate with the digital. Hansen will later go on to link that bodily response to Gilles Deleuze's cinematic concept of "any-space-whatever" and Marc Augé's "non-place," elaborating his thesis that digital art catalyzes a hapticity that both supplements the incommensurability of the digital world with human sensation, and at the same time creates an originary space in the body itself.[6]

It is tempting to see in Hansen's discussion of the digital distortion of the visual a direct analogy both to the auditory logic of the *Stretch* and the imaginative flux of Bachelard. By manipulating the duration of the recording, but otherwise leaving the pitches and timbres intact, the *Stretch* allows us to listen to sounds that are, outside of a computer, physically impossible to produce and thus otherwise acoustically impossible to hear in the first place. When, in the *Stretch*, we listen to the full chorus sustaining single notes for minutes on end, we cannot help but become aware of the regularity of own respiration against the sound of human voices vibrating in a breathless digital beyond. We feel the incommensurability of those two worlds *in* our bodies, and even as our mind explains to us the 1/22 ratio of the audio stretching, our body hovers in a corporeal elsewhere of embodied listening. At times, it is unsettling, uncanny; at other times, it approaches a kind of sublimity, the kind of transcendent reflexivity that Bachelard describes. But the sheer scale of the piece would seem to complicate Hansen's aesthetic of digital embodiment because the duration is beyond the span of human attention and indeed beyond the limits of what the human body can endure.[7] Built into the very structure of a 24-hour duration are moments of inattention, distraction—not to mention the need for food and sleep—that are at odds with the relentless physical and aesthetic continuity that the installation seems to demand. A theory of affective embodiment can only exclude the real bodily needs that are the necessary obverse of a technical process of 24-hour duration.

Searching for the *9 Beet Stretch* online, one will encounter images or videos shared on social media of the various performances of the piece over the past dozen years. And among those images, one will invariably see young people in sleeping bags, on couches, reclining with blankets and pillows. Because no one wants to stay awake for 24 hours, not even when they attend a 24-hour music installation. Yet following Crary's argument in *24/7*, which at the outset invokes military research into sleeplessness as well as torture techniques of sleep deprivation, one cannot help but think of some of the other great 24/7 "sound installations" of the early twenty-first century, namely US military bases and various "dark sites" hidden around the globe where 24/7 high-intensity lighting and loud amplified music are the conditions of an excruciating reality of enhanced interrogation (Cusick 2008). Such sites are, for Crary, both extreme and exemplary of "the expanding, non-stop life-world of 21st century capitalism," the temporality of which "can be characterized as the generalized inscription of human life into duration without breaks, defined by a principle of continuous function. It is a time that no longer passes, beyond clock time" (Crary 2013: 8). Crary further formulates the temporality of 24/7 as "time without time, sequence or recurrence" and "non-time" (29). What is at stake is the unprecedented intrusion of the incessant access and availability afforded by digital technologies into the social, intimate, and bodily dimensions of contemporary life. For Crary, digital technology transforms the traditional forms of alienation associated with industrial capitalism into a round-the-clock environment that seeks to unmoor the individual from any other rhythm (even that of day and night) that does not beat at the same pace as global capitalism. The result is a new version of modernist alienation in which the individual, completely cut off from any meaningful social context, must manage his own alienation through the technological apparatuses, which have assumed dominance over the social.

Visually, and with an eye to the round-the-clock illumination of US torture facilities, Crary argues that a 24/7 logic eliminates shadow and light, texture, ambiguity. It introduces a constant glare in which we all live: "Glare here is not a phenomenon of literal brightness, but rather of the uninterrupted harshness of monotonous stimulation in which a larger range of responsive capacities are frozen or neutralized" (34). Glare is thus the enemy of indeterminacy in that its effects amount to the equalization, reduction and foreclosure of responsiveness, exactly the opposite of what someone like John Cage surely meant by his open-ended durational processes. Crary's notion of a constant glare, beyond the rhythm of day and night, is crucial for the sensory, metaphorical and (ultimately) political significance of 24/7. It is intrusive, homogenizing, paralyzing, and it reveals a chilling continuity between the political infrastructure of round-the-clock torture and the economic infrastructure of digital capitalism. Translating Crary's glare into sonic terms— and imagining what it might be like to listen to "Enter Sandman" or the Sesame Street theme at high volume for days on end— we might begin to hear the din of a 24/7 world.

According to Crary's thesis, I have to admit that there is contradiction at work in my notion of transductive listening (and arguably, in the *Beet Stretch* itself), namely the fantasy that you can separate a technical process from the technological, and thus geopolitical infrastructures that make it possible. In her discussion of the militarized use of 24/7 music as torture, Cusick (2008: 18) points out that "[e]very amplified sound in the camps, and therefore every bit of music, *is* the United States' transformation of the energy in Middle Eastern oil into violent, violating sonic energy aimed directly at the people whose land yielded that oil" (emphasis in the original). In other words, it is ultimately impossible to think of 24/7 music(al torture) separately from the exploitation of the resources that sustain it because the music *is* that very exploitation. The geosonic infrastructure of music(al torture) is literally geopolitical in this instance. As much as transduction blurs the distinction between an object of listening (i.e., sounds, Beethoven) and the process of remediation itself, it can also reveal the structures of global capitalism in which digital technologies are ultimately embedded. Geosonically speaking: transductive listening is itself *determined* in the last instance by the economic, that is, by the 24/7 earth-extractive logic of late capitalism. The elsewhere in which we are situated by the *Stretch* is necessarily also the *everywhere*, the total environment of global capitalism. Likewise, when Mark Hansen develops a notion of embodiment in response to the alienness that digitally distorted works like *skulls* impose on human perception, what is missing is the larger structure of technological alienation that such artworks derive from, even if they attempt to mirror it back critically. Technical processes can be separated from the larger technological apparatuses that dominate social life only at the risk of creating a fetishizing fascination (e.g., an experience of immersion) that obscures the banality and violence of their operation. Holbein at least had the sense to project an anamorphic skull—that is, a cipher of violence and human finitude—as the foreground and frame of the tools of world domination (globes, sundials, astrolabes) depicted in the painting.

Looking back over the multiple senses of indeterminacy I have employed in this chapter, I characterize their interrelation as follows: if the indeterminacy of *4′33″* revealed the hidden musical continuity between sound and "silence" and if *0′00″* sought to uncover a hidden aesthetic continuity between art and life, then the 24/7 logic of the *9 Beet Stretch* discloses a continuity between life and digital technology whose sheer possibility is already implicated in the latter's necessary domination over the former. The aesthetic suspension of time accomplished by the 24/7 duration of the *9 Beet Stretch* (and other cyclical, continuously streaming, 24/7 digital works) is in fact not incompatible with the scale of human experience, but on the contrary constitutes precisely the ceaseless, (a)temporal environment of global capitalism.

There is a philosophical precursor to Crary's argument that draws together time, technology and musical aesthetics into a sustained critique of modern capitalism, namely Theodor Adorno. In moving to a conclusion, I want to make explicit the Adornian/Frankfurt School underpinnings of Crary's sweeping critique[8] by first turning to Adorno's ([1949] 2019) prescient formulations of 24/7 time in his *Philosophy of New Music*, and then returning to the dramatic caesura from Beethoven's 9th that initiated my analysis. According to Adorno, twentieth-century music had already given up on time with the debut of Stravinsky's *Rite of Spring* in 1913, a work which heralded a new compositional style that sought to replace the experience of musical time by a technique of musical spatialization. For Adorno, musical spatialization involves a series of structural shifts in the art of composition: repetition replaces development; modality replaces tonality; form is suspended in favor of excessive coloration; and counterpoint gives way to a diffuse musical atmosphere. Stravinsky is not original in this regard, for he merely radicalizes a tendency toward spatial composition already seen in Wagner (where the concept of the musical drama predominates over musical time) and in Debussy (where atmosphere absorbs temporal progress into an impressionistic stasis). Stravinsky intensifies this trend by subordinating all other musical effects to the sole aim of spatializing musical time. Adorno writes, "One trick defines every manipulation of form in Stravinsky and is soon used to exhaustion: Time is suspended, as if in a circus scene, and complexes of time are presented as if they were spatial" (142–3). Practically, this means that, in Stravinsky, we hear a music that has "renounced all possible means for the production of time-relationships—transition, intensification, the distinction between the field of tension and the field of release, further of exposition and continuation, and of question and answer" (Adorno [1949] 2019: 143). The montage-like process of composing music in rhythmic and melodic complexes without any organic connection or development has the result that "the musical continuum of time itself is dissociated," abandoning "the dialectical confrontation with music's temporal progression" (138).

For a thinker like Adorno, for whom art, society, history and technology are in permanent dialectical interaction, this is no mere stylistic deviation. The musical abandonment of time expresses the profound impasse of culture in the era of high capitalism and may be linked to the—at the time of his writing (the late 1940s)—recently emerged and even more aggressive forms of domination associated with totalitarianism and fascism. Commenting in the same essay on Wagner's tendency toward spatialization, Adorno writes, "The suspension of musical time consciousness corresponds to the entire history of the bourgeoisie, which, no longer seeing anything in front of itself, denies the process of history itself and seeks its own utopia through the revocation of time in space" (140). Indeed, the fantasy of a spatialization of (musical) time could only emerge out of a social class that believes it has itself triumphed over history, that sees time as a given and space as the only dimension in which it exists, spread out before it like some open territory waiting to be claimed. Such a perspective helps us to contextualize the notion of the soundscape as a dominant-class and settler-colonialist view: the immersion in the soundscape depends on a prior appropriation of territory that allows it to appear imaginatively as an environment of pure sonic availability to a certain kind of listener. Yet the flipside of that triumph is the paradoxical domination of the middle class by the very capitalist processes that ostensibly empower it. And that domination extends most sinisterly into the domain of time, for Adorno seems to suggest that the most dangerous outcome of bourgeois ideology is an individual who cannot think for himself because he has no time to think for himself.

In an earlier text, the famous chapter on the culture industry from the *Dialectic of Enlightenment*, Adorno and Max Horkheimer write, "The bourgeois whose existence is split into a business and a private life, whose private life is split into keeping up his public image and intimacy, whose intimacy is split into the surly partnership of marriage and the bitter comfort of being quite alone, is already virtually a Nazi, replete with both enthusiasm and abuse" (1972: 155). The individual who is forced to subdivide his life into increasingly smaller units and "time-manage" his own alienation has in fact forfeited his individuality to a system that dictates his actions at all times because it has eradicated the spontaneity of "unscheduled," free thought. Even when he is alone, the bourgeois is still on the clock, his solitude sterile and unproductive because it beats time to a rhythm prescribed by a system he obeys without question. Turning back to Adorno's essay on Stravinsky, we note an astonishing link between the reified, proto-fascist consciousness of the culture industry and the rarefied products of European art music, for Stravinsky's music produces almost exactly the same attitude of distraction and submission. Adorno insists that Stravinsky's abandonment of musical time must be understood not so much as aesthetically motivated, but also as the result of "the pressure of a system whose irrational superiority over everything subjected to it maintains itself exclusively on the basis of estranging people from the effort of thinking and reducing them to mere centers of reaction, to monads of conditioned reflexes" (Adorno [1949] 2019: 146). The listener of Stravinsky finds himself in the same position as the consumer of the products of the culture industry in that all his reactions are calculated in advance. The music, spatially conceived, does not allow for a becoming in time, for the free play of subjective thought in dialectical tension with the objective act of listening. His music produces a passive listener, at best, and taking that passivity to a dialectical extreme, Adorno writes: "Stravinsky's *fabula docet* is versatile compliancy and obstinate obedience, the model of that authoritarian character that today proliferates on all sides" (ibid.).

Adorno's treatment of Stravinsky is certainly heavy-handed and polemically charged, yet what underlies it is a distinction between two modes of listening which, I argue, will help us to understand what the *9 Beet Stretch* does with time, and, perhaps not by chance, with Beethoven. Adorno names an "expressive-dynamic" and a "rhythmical-spatial" mode of listening and suggests that the two modes interact throughout the history of Western music, with either one predominant at a given time:

> The former [expressive-dynamic] has its source in singing; it aims at surmounting time through its fulfillment and, in its supreme manifestations, inverts the heterogeneous movement of time as a force of the musical process. The other type obeys the beat of the drum, intent on the articulation of time through its division into equal quantities that virtually abrogate and spatialize time. The two types of listening diverge by virtue of social alienation, which tears apart subject and object, separated by force of that social alienation which separates subject and object. [...] The idea of great music consisted in a reciprocal interpenetration of these two types of listening and compositional categories that conformed to them.
>
> Adorno (1949) 2019: 144

Adorno defines listening not so much in relation to our perception of musical form unfolding in time as to our understanding of music's interaction with time itself. In this regard, he estimates, in the *Philosophy of New Music*, Beethoven's 7th Symphony most highly as a work, both in its formal construction and at the level of his dual typology of listening, which offers a musical and philosophical synthesis of expression/rhythm, song/dance, and subject/object. Yet, in passing, he remarks that the late works of Beethoven (which would include the 9th Symphony), renounce precisely the dialectical synthesis attained in the 7th: "[Beethoven] himself in his late phase surrendered this paradoxical unity and, as the highest truth of his music, allowed the absence of reconciliation between the two categories to obtrude baldly and eloquently" (145). What is crucial for the polemic is the contrast between Stravinsky's cheap, pseudo-dialectical spatialization of musical time as montage, and the more profound fragmentation of time in the late Beethoven.

Though Adorno remains focused on Stravinsky for the remainder of his *Philosophy of New Music*, the short essay, "Late Style in Beethoven" (2002), dilates upon the breakdown of dialectical time in Beethoven. Here Adorno argues that "late style" in Beethoven and other artists, is characterized by a kind of abandonment of the subjective shaping of artistic material that we normally think of as "style." What emerge are not smooth interpenetrations of form and content, style and idea, but rather radical breaks, which interrupt the work of the artist like death itself:

> The caesuras, the sudden discontinuities that more than anything else characterize the very late Beethoven, are those moments of breaking away; the work is silent at the instant when it is left behind, and turns its emptiness outward. Not until then does the next fragment attach itself, transfixed by the spell of subjectivity breaking loose and conjoined for better or worse with whatever preceded it; for the mystery is between them, and it cannot be invoked otherwise than in the figure they create together. [. . .] He does not bring about their harmonious synthesis. As the power of dissociation, he tears them apart in time, in order, perhaps, to preserve them for the eternal.
>
> Adorno 2002: 567

Bracketing for the present context the difficulty of what Adorno implies in the late work's imbrication of form, matter, life and death, we can at least say that the relentless continuity of a 24/7 work like the *9 Beet Stretch* gestures toward a world that radically denies the possibility of fragmentation, cessation, and release from its near unendurable duration. The ultimate indeterminacy of listening would not be sought in Cagean continuities, Hansen's embodiment, or Bachelard's imaginative elsewhere, but perhaps in that exemplary caesura from the 4th movement of the 9th. As the pause slowly unfurls into and interrupts the soundscape of the *Stretch*, the sonic gap transduces the earth that is otherwise inaudible in 24/7 capitalism.

4

Now You See It . . .

Hearing Colors and Picturing Soundscapes

Soundscapes posit mediated relations between an external environment and a listening interiority. Listening through soundscapes is a practice of critical attention that activates the outside world, the inner sensorium, and the media that connect them as sites of transduction and transformation. The previous two chapters emphasized an outside-in movement, in which complex geologic environments (the earth as a seismic field in Chapter 2 and digital capitalism as an earth-scale system in Chapter 3) are transduced as soundscapes by technical and aesthetic listening. This chapter takes an inside-out approach, exploring how an artist's unique embodied technical and sensory abilities re-present the environment as a soundscape for audiences, and how art affords critical opportunities to rethink (rather than reify or essentialize) our complexly mediated connection to the soundscape.

I begin with the art and activism of Catalan-raised, London-based cyborg artist Neil Harbisson. In the first part of the chapter, I critically situate Harbisson's self-made and self-promoted status as a cyborg within the discourse of transhumanism and the broader enhancement rhetoric of contemporary neoliberalism. I suggest that Harbisson follows the transhumanists in appropriating discourses around gender, sexuality, and ability to legitimize a neoliberal rhetoric of bodily enhancement through technology. Harbisson's cyborgism in fact hinges on his rare eye condition of total color-blindness, which Harbisson transforms into an "artificial sense" through a prosthetic implant that sonifies color in his head. But in thus sonifying color into a customized cyborg soundscape, Harbisson also effectively silences questions around living and creating with disability, as well as how to explore inclusive forms of intersensoriality in aesthetic experience. Ultimately, I suggest that Harbisson's cyborg-sonified soundscape remains his private property, stuck within a neoliberal framework that understands embodiment (and its creative products) as a kind of empowered ownership.

In the second and third parts of the chapter, I attend more closely to the techniques and processes that underlie Harbisson's creative practice, considering them critically alongside transductive phenomenologies of embodiment and technology (from Adrian Mackenzie and Mark Hansen). I argue that Harbisson's work has the critical potential to shift the location of cyborg-ness from techno-bodily prosthesis to a model of networking and an aesthetic of the soundscape: the body is (re)connected to the environment by technical processes of transduction and transposition that *could* amplify the indeterminacy of being (in) a body, rather than merely augmenting the body's experience of reality. Stelarc's *Ping Body* (as analyzed by Mackenzie) is a key example of such an indeterminate cyborg art practice. I argue, by contrast, that Harbisson misses the productive opportunity to use his art to communicate a collective sense of embodied indeterminacy, instead retrenching cyborgness in individualist, essentialist, and techno-determinist conceptions of color, sound, and nature. In the concluding fourth part, I consider the work of Deaf artist Christine Sun Kim, who likewise uses sound to activate a seemingly unavailable sensory modality, but in a completely different way. Kim folds listening into an intersensorial aesthetic that redistributes how deaf and hearing audiences experience sound in social environments.[1] She crafts idiosyncratic, pictorial soundscapes which situate embodiment, affectivity, and the senses in an indeterminate, creative interaction with sonic and social space, resonating across a multisensory aesthetics and an anti-audist social field.[2]

Part I. Old Cyborgs in New Bodies

Neil Harbisson is a self-styled cyborg artist and claims to be the world's first officially recognized cyborg person, owing to the Wi-Fi-enabled antenna that has been permanently attached to his skull since 2004.[3] Harbisson was born with a rare eye condition called achromatopsia, a kind of total color-blindness whereby he can accurately perceive the intensity and saturation of light, but otherwise sees the world in total grayscale. The antenna (which he sometimes refers to as an "eyeborg") detects the electromagnetic waves that human eyes usually register as color and converts those waves into sounds which vibrate in Harbisson's skull and which he thus hears through bone conduction. A prototype of the antenna was externally attached to the back of his head and powered by a laptop computer, but a subsequent version has been surgically drilled directly into his skull and carries out the light-sound conversion with its own built-in electronics. And so, through the implant, Harbisson can essentially "hear" color. The particular tones he hears at any given time depend on the frequency of the light transposed onto a "sonochromatic" scale (which I discuss in more detail in Part II), and the volume of the tone correlates to the measured saturation or intensity of the color. Infrared and ultraviolet frequencies, otherwise invisible to humans, can be detected and sonified by Harbisson's antenna. The antenna also allows Harbisson to receive phone calls and listen to music via satellite uplink, and he can also "hear" videos and photos sent by select friends on five continents, even in his sleep. More recently, he has collaborated with NASA to link his antenna to solar cameras orbiting the earth. Harbisson insists that he has invented a new bodily sense, a kind of synaesthesia or, as he calls it, sonochromatism, and that the sonified perception of color has become completely integrated into his everyday life, and his perception of people, faces, cities, and natural environments.

Bracketing for now the technical and aesthetic affordances of the antenna, I want to dwell on Harbisson's cyberactivism. For it is arguably less as an artist and more as an activist that Harbisson wants to make his mark, and all his cyborg art can be seen as a promotion, indeed a self-promotion, for a particular kind of identity (and identity politics). In 2010, Harbisson co-founded the Cyborg Foundation to: (1) defend cyborg rights; (2) promote cyborgism as an art movement; and (3) support people who want to become cyborgs.[4] An offshoot of the Cyborg Foundation is the Transpecies Society (launched in late 2017), with similar interests in raising awareness and creating collaborative and social spaces for persons with technologically designed nonhuman identities.[5] A promotional video from (the November 2017 version of) the front page of the Cyborg Foundation's website announces that "We have the freedom to merge technology with ourselves, honouring our transpecies origins, reconnecting with nature and creating a more balanced relationship between us and the universe" and further down the page we read "It's time for transpecies to come out of the closet." And indeed, the first step to becoming a cyborg is "IDENTIFY," by which the site means to identify or choose the particular capability or body part one would like to modify, but which also implies choosing "cyborg" as an identity.

These snippets from the website already give us some orientation toward Harbisson's cyborgism as a stable identity category premised on the unproblematic union of technology, information, and the body. Such an identity entails repressing at least a century of cultural history that has consistently presented the cyborg as a violent, uncanny figure living out the incommensurabilities and violences inflicted on human bodies and senses by capitalism.[6] In the place of this cultural history, Harbisson's cyborg substitutes a transhumanist and transpecies rhetoric that reinstalls an autonomous liberal subject at the helm of a critically unnuanced concept of technology.

The notion of the transpecies is rapidly evolving and subject to widely varying interpretations,[7] but for Harbisson it seems to suggest a deep, if somewhat vague link, through morphology and evolution, between human and other animal bodies. In technologically modifying the senses, Harbisson's cyborg does not become "super-human," but rather reveals affinities between the human and other animal forms, thus opening up a transpecies terrain beyond and between conventional human-animal relations. Harbisson has already indicated as much about his own antenna, describing how he initially hit upon the idea of a chromosonifying prosthetic: "I kept searching for ways to create a new, independent body part, and I found what I was looking for in nature. Many animals, especially insects, are equipped with antennae. So why couldn't I be?" (2015a, my translation). For Harbisson, however, such trans-species affinities can only be experienced by technological modification of the human body, specifically the development of new sensory capabilities. Those affinities are not revealed by building new kinds of relationships with animals or enacting a different kind of ethics regarding the treatment and rights of animals, but rather only by an appropriative technological act: "Technology as a tool estranges us from nature. If you yourself become technology, you can use your expanded senses to feel connected to other animal species and get to know nature from another angle" (2015a, my translation).

When we read on the Transpecies website that it is time for transpecies to "come out of the closet," the invitation is clear to view cyborgism as a marginalized identity analogous to queer and transgender identities. When asked how many people could currently be called cyborgs, Harbisson responded, "That question's just as impossible to answer as a question like how many men feel they are really women. In both cases it's a matter of identity" (2015a, my translation). He further develops this identity parallel, linking the discrimination, repression, and violence endured by LGBTQ2S+ people with the political struggles and social antagonism faced by cyborgs. The rhetoric of Harbisson's website is clearly trying to establish the legitimacy and indeed urgency of cyborg politics by analogy to LGBTQ2S+ politics, suggesting that a person's right to construct their own gender and sexual identity (even, and especially, in cases of sex-reassignment surgery and other forms of gender-affirming care) is analogous to a person's right to design and modify their sensorium through surgery, implantation, neural interfaces, etc.

On a legal and/or ethical plane, these rights to fluid identities and body modification may indeed be analogous, but it is crucial, I think, to raise the objection that, socially, the cyborg and the LGBTQ2S+ may not be so commensurable after all. In an era when even the most basic aspects of social, professional, and domestic life require a hand-held, networked device and when digital technologies have so radically permeated and rewritten the functioning of markets, institutions, and administrative life, it seems absurd to claim there are *any* forces out there (social, political, or otherwise) bent on repressing the individual's personal relationship with technology. On the contrary! One could argue that the coercion to be a consumer of social media and the ever-proliferating array of customizable digital technologies and services is as strong as the coercive, heteronormative mechanisms that still continue to regulate gender and sexuality. And when real homophobia and the very real risks and dangers assumed by those with LGBTQ2S+ identities are likened to the discriminations cyborgs may experience because of their "out" relationship with "technology," it is clear, I think, that not all marginalizations are commensurable. From the perspective of Harbisson's cyborgism, there is an assumption of equivalence between how a body relates to gender and how a body relates to technology. In this view, gender fluidity does not pertain to a domain of bodily life that is separate or distinct from technology; rather the very concept of gender fluidity would seem to entail (the possibility of) technological body modification.

Yet what is disavowed in positing the cyborg analogy of gender and technology is the question of disability. Harbisson self-identifies as a cyborg, a transpecies, an artist merging body and technology to create new senses and experiences; yet he also has a potentially disabling condition, the congenital color-blindness whereby he cannot experience the visual sensation of color. As much as his antenna enables a synesthetic transformation, it is also, on some level, a means of overcoming or compensating for his differently abled body (but not identifying as disabled). In this sense, the cyborg configuration of body/*ability*/technology is potentially more compelling than body/*gender*/technology. Moreover, given the conspicuous nature of Harbisson's antenna, the kinds of vulnerabilities and antagonisms he is routinely subject to in his daily life are perhaps more closely aligned with the experiences of disabled, rather than LGBTQ2S+ individuals. At a 2011 demonstration in Catalonia, Harbisson's antenna—and thus his body—was damaged by police, who suspected Harbisson was filming them. Here Harbisson was at greater risk and an easier target for police mistreatment solely because of the conspicuousness and vulnerability of his visibly altered body, not his gender or sexuality.

Whether deliberately or not, Harbisson appears to downplay, if not exclude altogether, the experience of disability from the rhetoric of his self-presentation. This is perhaps less a question of (dis)identification with disability and disability movements than it is a matter of what Mack Hagood calls audibility (2017). Writing about tinnitus, Hagood argues that disabilities require mediation to become *legible* diagnostically, *visible* as part of a distinct community, and *audible* as a meaningful fit between a person's sense of self and a broader public discourse: "Audibility means being able to include one's physical experience in an interior monologue that makes sense within public discourse. It also means being able to express one's experience in terms that will be sensible to others, whether through dialogue or activist invective. Without audibility, there is no creation of disabled publics and thus less agency for disabled people" (2017: 317). Disability, for Hagood, is thus triply mediated by (1) medical discourses and diagnostic instrumentation; (2) social rhetorics of the visualization of distinctive bodies; and (3) communal discourses with which a person identifies by hearing and speaking himself into a particular identity position—a recognition and identification through audibility. The case of tinnitus as a disability, according to Hagood, is problematically a question of audibility alone, since the medical diagnosis and the visualization of a tinnitus-suffering body do not readily accompany the condition's audibility as both an embodied experience *and* a communal narrative with which tinnitus sufferers can identify. Harbisson inverts the tinnitus example of mediated disability: his condition is medically legible and his prosthetic renders him visible as a distinctive body; yet in not engaging the audibility of disability, that is, of hearing/speaking himself into a socially recognizable identity position as disabled, Harbisson ambiguously turns away from a politics of disability. Precisely in trying to sonify color, Harbisson renders disability inaudible.

Numerous disability scholars have remarked on the critical potential of disability to expose the mechanisms of ableist ideology (Siebers 2008); social and corporeal normativity (Davis 2013; Garland-Thomson 2017); the exercise of Foucauldian biopower on individual bodies (Tremain 2015); the neoliberal flexibility that increasingly constrains individual and collective agency in the twenty-first century (McRuer 2006; Hagood 2019); and the diminishment of nonhuman materiality associated with neoliberal subjectivity (Mitchell, Antebi, and Snyder 2019). In bypassing the critical potential of disability, Harbisson does not address the normativity and ableism that these scholars would associate with the techno-deterministic, neoliberal subject position that he embodies. Tobin Siebers writes that "the emerging field of disability studies defines disability not as an individual defect, but as the product of social injustice, one that requires not the cure or elimination of the defective person, but significant changes in the social and built environment" (2008: 2). Rather than conceive of his achromatopsia as a means of questioning the normative assumptions around the experience of color and the ways it is codified in institutional and daily life; in built and natural environments; and in the conventions of the visual arts and aesthetics, Harbisson acknowledges his condition only insofar as he corrects it via an all-powerful technology that posits enhancement as a new bodily norm. For Harbisson, the cyborg seems to silence, or render inaudible, the complex embodiment and critical epistemology inherent to the experience of disability (Siebers 2008: 25) as well as the critical affinities between queer and crip identities and experiences (McRuer 2006).

Harbisson's casting of the cyborg as potentially closeted, entitled to the same kinds of rights and recognition as LGBTQ2S+ individuals, and his use of the term transpecies evokes the cyborg body of another trans-movement: transhumanism. Transhumanism is a kind of life-philosophy or vitalism for the twenty-first century that promotes the active transformation of the human body—and the human species—through technology. It may have started as a somewhat marginal, arguably new-agey cultural movement, but has increasingly become an area of serious academic and entrepreneurial interest.[8] Some of its key objectives include radical life extension, the uploading of human minds to computers and the creation of "substrate-independent minds," that is, minds that are not confined to a single human body, but can be distributed and synchronized across an ensemble of digital and corporeal platforms. Both transgenderism and disability are fundamental to the transhumanist concept of the body. Martine Rothblatt has argued that an individual's right to freedom of gender is "the gateway to *freedom of form* and to an explosion of human potential. First comes the realization that we are not bound by our sexual anatomy. Then comes the awakening that we are not limited by our anatomy at all" (Rothblatt 2013: 318). Rothblatt envisions a cyber citizenry of *persona creatus* [*sic*]: avatar-like entities living in and reproducing through a multiplicity of flesh and digital forms. That the transhumanist mind is deeper than bodily matter is an insight Rothblatt derives directly from transgenderism: "Consciousness will be free to flow beyond the confines of one's flesh body as gender is free to flow beyond the confines of one's flesh genital" (322).

Rothblatt is both a high-profile transhumanist (a multimillionaire CEO with interests in satellite communications, aviation, and biopharmaceuticals) and herself a transgender person. And yet, at the noncorporate, DIY end of the transhumanist spectrum, we find a similar link between transhumanism and the transgendering of the body. Tim Cannon, a DIY body hacker, runs a biotech start-up out of his basement on the outskirts of Pittsburgh. In an interview recounted by Mark O'Connell, he writes, "I'm trapped in this body. [. . .] But that's not just a religious idea, man. Ask anyone who's transgender. They'll tell you they're trapped in the wrong body. But me, I'm trapped in the wrong body because I'm trapped in *a* body. *All* bodies are the wrong body" (O'Connell 2017: 158). In Cannon's case, as in Rothblatt's, the specificity of the body's relationship to gender leads to a general claim about the relationship of all bodies to technology. Whereas Rothblatt's gendered body is a source of potential, Cannon's gendered body is a source of constriction, a trap. In either case, technology steps across the conceptual threshold of gender to alter or expand the body.

Disability enters the conceptual language of transhumanism through the notion of "morphological freedom," which Anders Sandberg defines as "an extension of one's right to one's body, not just self-ownership but also the right to modify oneself according to one's desires" (Sandberg 2013: 56). Sandberg's version of transhumanism is more traditionally liberal than Rothblatt's in that his rhetoric of creativity and self-expression relies on concepts of happiness, self-ownership, and personal autonomy. For Sandberg, a creative drive toward self-expression is to be preserved in a legal framework that guarantees freedom for all individuals to pursue their happiness to the fullest, without the interventions and interests of governments, health care and insurance industries, etc. For him, the right to morphological freedom amounts to a right to a differently abled body: "It should be noted that the disability movement has been a strong supporter of the right to determine one's body just for this reason. This seems to be a natural point of agreement between transhumanists and the disability movement [whose] postmodern critique of the normal body also supports the right to be differently bodied" (62). Note how Sandberg selectively invokes disability as a "right to be differently bodied," and not as a right to equal forms of social access and participation without stigma or prejudice. Recalling Siebers' quote above, disability is not an individual defect, but rather a product of an unjust social environment. Morphological freedom thus co-opts disability insofar as it bends the concept to fit its enhancement rhetoric: morphological freedom amounts to the freedom to buy expensive technological enhancements, not freedom of equal social access.[9]

There is a pseudo-queer and crypto-ableist logic at work in the vitalism of the transhuman body, which leaps to the following kinds of conclusions: because some (transgender) people experience their body as the "wrong" body and choose to modify it, therefore *all* bodies are inadequate and in need of modification; or, because some groups have been victims of oppression and marginalization for their bodies (e.g., through race, sex, gender or ability), *all* people (especially white, male transhumanists) are in some way victims of their bodies and deserve/need technological empowerment. There is a further danger of positing dubious equivalences: to link the historical specificity of, say, slavery or the subjugation of women to the quasi-universal experience of being "slaves to our genes" (More "Letter" 450) or incarcerated, as Rothblatt has it, in a "prison of sex" (324) is to simultaneously obliterate, appropriate, and tokenize the particular violence of, say, racism and misogyny in order to set up a false continuity between struggles for civil rights and struggles for transhumanist rights.[10] The struggles for civil rights took place historically (and still take place) in a racist and sexist society that was/is radically hostile to the rights of women and African Americans. The same simply cannot be said for the struggle around transhumanist and cyborg rights because the contemporary West is not fundamentally hostile to technology in ways even remotely commensurable to how the West has historically been hostile to women, peoples of African descent, and other marginalized groups. The absurdity of the part/whole totalizing logic that infuses the transhumanist concept of the body is seen in the movement's imperative to space exploration: not only is embodiment a prison, but indeed the whole terrestrial habitat is a confinement to be abolished by a hierarchical, quasi-imperialistic expansionism of consciousness that calls itself, quite tellingly and with palpable historical and cultural amnesia, colonization.[11]

Despite obvious affinities with transhumanist philosophies of technology and the body, Neil Harbisson does not explicitly describe himself as "transhuman" nor does the term appear on the websites of the Cyborg Foundation or the Transpecies Society. For that matter, Harbisson is not mentioned in any of the literature on and by self-identifying transhumanists that I have read. One point of distinction may be Harbisson's favoring of concepts like RR (revealed reality) and AS (artificial senses) in explicit contrast to transhumanist-friendly discourses of augmented or virtual reality, and AI (artificial intelligence) respectively. Yet the overlaps are clear. Harbisson's cyborgism shares with transhumanism an imperative to enhancement to express full ownership over the body and its senses, grounded in a techno-deterministic, neoliberal framework.

Part II. The Cyborg and the Soundscape

It is perhaps heavy-handed to fault Neil Harbisson for not taking the twentieth century (pre-)history of cyborgism into account in his art and activism. But what of the immediately preceding generation of performance and body artists who explicitly incorporate and perform cyborgian themes in their work, and enact key technical and philosophical problems of bodies and technologies associated with postmodern and posthuman discourses? Artists like Orlan, Eduardo Kac and Stelarc produced widely influential work in the 1990s and early 2000s that re-envisioned "classic" cyborgian themes in decidedly performative and postmodern terms, with a particular emphasis on the body's increasing entanglement with new media, digital infrastructures, and biotechnologies.[12] If the classic modernist version of the cyborg regarded the body as a kind of unity and interiority that was to be either monstrously disfigured or triumphantly transformed by an exterior force called technology, the posthuman version regards the body as a heterogeneous entity whose porous inside/outside boundaries are being constitutively (re)drawn by technologies from the very outset.[13] Yet, as much as such discourses illuminate the history of the human body in and through technical systems, they also do little to dispel the fundamental strangeness and contingency of the lived experience of technology. That experience of the body—as technological interface, transducer, and spectacle—is precisely what artists like Orlan, Kac, and Stelarc attempt to capture and perform in their work. And as immediate precursors to a "cyborg artist" like Harbisson, their work offers a critical point of comparison for understanding what kind of experience of technology is embodied in Harbisson's art.

Harbisson's antenna plays on the convertibility of sound and color as different forms of energy that can be described or processed as waves. Light is wave-form energy, and the color spectrum is characterized by the range of frequencies of light waves that interact with our eyes and brains to produce the sensation of color. Likewise, sound is a differential vibratory energy that propagates in waves, which, within a given frequency range, resonate in the human body and brain to produce the sensation of hearing. Even if the mechanisms of these operations are far more complex (and they certainly are) than this crude sketch, it remains a fact that color and sound are both vibratory energies transduced into sensation by the human brain. Harbisson's antenna effects a transposition of energy from one range to another along the frequency axis, suggesting a quantitative continuity between vision and hearing, rather than an embodied or experiential distinction.

Harbisson has developed a Sonochromatic Music Scale that maps twelve colors of the visual spectrum (from red to violet) onto twelve tones of the Western musical scale (from F to E, respectively). Although the sonochromatic scale and the antenna allow for up to 360 microtones (to transpose 360 degrees of the color wheel), it basically works out that the color red corresponds to the note F; green to A; blue to C-sharp, etc.[14] Harbisson insists that the transposition is "objective," based on cycles and intervals that are intrinsic to the quantities of energy that underlie the color wheel and the Western musical scale. Thus, in his view, the sonochromatic scale avoids the arbitrariness of other diagnosed synesthetes who experience color as a sound (and vice versa) in their individual, idiosyncratic ways. Harbisson, by contrast, understands himself as approaching a general formula or standard by which color may be transposed to sound. In various interviews and talks, Harbisson refers to the long history in Western science of studying sound and color together, mentioning in passing the work of Isaac Newton on the subject. Newton was convinced that light had a cyclicality analogous to the musical octave, and devised a color scale accordingly. Proceeding from "deepest scarlet" to its supposed octave "deepest violet," Newton plotted the remaining colors along mathematical intervals that matched the Dorian mode. Yet as Peter Pesic has shown, these intervals were not born out by experiment nor theoretical calculation, and Newton was forced either to omit colors to match the scale, or to recalculate the intervals of the wavelengths of the visible spectrum, essentially pretending that the ratio of red to indigo (400nm:700nm) was actually an octave (1:2).[15]

The premise of a seamless quantitative transposability between sound and color breaks down both theoretically and practically. The math does not work, and thus the idea of an intrinsic, quantitative mapping of a frequency range of color (e.g., red) onto a frequency of range of sound (e.g., F above middle C) turns out to be an arbitrary sonification, rather than an objective conversion of sense data. In fact, precisely in conceiving of the operation of the senses as an informational process (in which the sensory data can simply be plugged into different software applications), Harbisson downplays sensation as an embodied, environmentally situated process. He subscribes to what Jonathan Sterne has called the "audiovisual litany," that is, a series of biases and conventions by which scholars unwittingly differentiate seeing from hearing without recognizing that they belong to an intersensorial ensemble of embodied sensation (Sterne 2012: 9–10). W. J. T. Mitchell identifies a similar limitation in aesthetic discourse, in which the concept of medium specificity has historically obscured the more complex perceptual and medial reality that "*all media are mixed media*" (2005: 260, emphasis in original). Rather than exploring the multiple sensory, perceptual, semiotic, and medial elements that compose "a" bodily sense in the first place, Harbisson proceeds from a normative cultural preconception of discrete senses that are technically interchangeable. By insisting that color and sound are quantitatively the same, Harbisson problematically reifies them as absolutely distinct sensory modalities.

A number of Harbisson's artworks and performances depend upon the putative accuracy of this transposition. At the beginning of his TED talk we see Harbisson holding up different colored cloths up to his eyeborg. The "corresponding" sine tone is piped in over the PA system, and Harbisson concludes that this cloth is purple, that cloth is green, and so forth. As presumably non-color-blind audience members, we cannot help but verify the accuracy of his transpositions and are invited to regard with wonder or insight that a purple cloth "really" is purple. There is a kind of tautology at work here that recalls Magritte's famous gesture in *The Treachery of Images* (1928) of painting a pipe with the caption "Ceci n'est pas une pipe." Only Harbisson seems to want to gloss over Magritte's lesson about inexorable gap between representation and things. Indeed, the only caption we might apply to Harbisson's demonstration is, "This purple cloth is/sounds like this purple cloth." An arbitrary substitution or detour through sound brings us back to the self-evident starting point, and that cyborg feat is no more than a reappearing rabbit in a magician's hat that was there all along. The antenna works like Cage's microphone in the anechoic chamber (discussed in Chapter 1), only Harbisson's work does not entail the challenge to our attention that Cage's ostensibly does.

In his video/sound piece "Sound Portraits" Harbisson takes a range of color readings of an individual's face—Harbisson prefers to work with celebrities and artists like then Prince Charles, Moby, Philip Glass, Nicole Kidman, and others

(2010). He edits together the composite color-sound values of hair, eyes, skin, lips (etc.) into a kind of musical chord. Harbisson then further edits different portraits together into a kind of short piece of music. Each portrait plays for a couple of 4/4 bars, pulsed to the same rhythm each time, with a faint metronome-like beats in the background. The sounds are sine tones, yet bristling with dissonance from the overlaid microtones. The portraits are arranged in contrasting pairs according to pitch: we hear two bars of the high-pitched and particularly jarring face of Daniel Radcliffe, followed by the lower and mellower Gael García Bernal, et al. Altogether there are 13 portraits in the 52 second mix, and each portrait is accompanied by its subject's name as a white caption on the black screen. Without the caption, we would have no other way to link face to sound, and there is, arguably, nothing especially revealing or unique about the sounds we hear as portraits, which is why the choice of celebrity subjects is decisive. They solidify the "accurate" and, indeed, auratic effect of the portrait, inviting us to believe in the intrinsic or essential relation of the sound to the absent referent in the same way that we were invited to verify the accuracy of Harbisson's conclusions about the colors of pieces of cloth. We hear a resemblance that is neither real, nor virtual—indeed, it is not even a resemblance, but merely a sonic transposition of that most banal of images: a celebrity headshot. The sound portraits suture a name to a sound in the performance of a pseudo-indexicality that covers up for the inevitable randomness of the connections. The portraits revisit Cagean ideas about microphone placement and the redirection of aesthetic attention, only here in the form of light entertainment tied to celebrity culture, with avant-garde framing.

Harbisson also builds musical compositions out of sound-color correspondences. In a piece jokingly titled "Sock Sonata No. 1," Harbisson plays the sounds of different colored socks and weaves the tones into a kind of electronic dance music composition (2013). He begins by holding socks one at a time to his eyeborg and letting a long sine tone resound throughout the hall. (Again, the sine tone is piped over the PA system from Harbisson's computer, which appears to be synced to the eyeborg.) After a couple of minutes of single, long tones, Harbisson puts his socks down, and a green box begins to flicker on an overhead screen like a kind of cursor. We hear a pulsed sine tone in sync with the flickering green box, which is soon joined by a pale blue box and a red box, and two new pulsing tones. The three tones/colors enter a repetitive pattern, and, with a whisper of beats and clicks in the background, a kind of crude rhythmic figure emerges: pre-programmed beats from Harbisson's laptop. Every now and then a high-pitched tone and a purple box add a harmonic and visual accent to the beat. We are now in the presence of an audiovisual piece that is part Atari console, part minimalist EDM. Harbisson then begins deliberately picking up one sock after another, cuing in long tones in various consonance and dissonance with the looping triad. After a couple of minutes, the triad resolves to the single pulsed "green" tone from the beginning of the piece. Harbisson holds a white card to his eyeborg with one hand to silence it, while tapping keys on the laptop with the other, and as the piece comes to a close, the artist bows to the audience's applause. The performance has a structure of the performed banality of Cage's *0′00″*, only the seeming turn away from conventional notions of performance and audience is reversed and reinforced in the rituals of bowing and applause.

Beyond programmed and interactive beats, Harbisson has also staged a site-specific "colour-conducted concert" for choir and string quartet (2015). The piece relies on the color palette of the mosaics and intricately ornamented ceiling of the Palau de la Música Catalana in Barcelona, where the piece was performed. Harbisson created a chord progression based on the colors he heard in the space and scored the progression by training a choir and string quartet to sing and play precise pitches in response to different colored lights. At the performance, various groups within the choir are illuminated at different times with different colored lights, vocalizing their corresponding pitches to produce harmonies. Likewise, the quartet plays tremolo on different notes indicated by separate lighting cues. Harbisson, dressed in white tails, conducts the ensemble from a music stand at center stage. It is hard to tell the duration of the piece and get an overall musical sense of it from the heavily edited video; but after the crescendo of its conclusion, Harbisson is moved to tears as he delivers a short speech on this experiment in which musicians "perform colour" for the first time. The rhetoric of the performance combined with the compression of our attention back on the performer (instead of its expansion outward into post-Cagean indeterminacy) catalyzes a profound affective response from Harbisson, infusing his work as a technologist of color with the pathos of a Romantic composer.

Harbisson's cyborg body and artistic persona ground the otherwise arbitrary color-to-sound conversion into artefacts that resemble works of art, guiding our attention away from the technical processes that underlie them and toward the essential properties they promise to reveal about color, sound, and sensation. Throughout such artworks, the audience is not directly challenged to imagine, in a sustained manner, what it would be like to experience a world without color, or to experience a world in which such a taken-for-granted sensory element is only accessible in altered form via an infrastructure of technical transduction. Rather, the artworks present a seamless translation of color to sound by creating finished products (e.g., a piece of music, or a sound-portrait) that make that infrastructure disappear.

Harbisson, for all the interviews, performances, and lectures he has given, reveals very little of the actual technical specifications and operation of the antenna. Aside from the basic aspects of the design and structure, he does not comment on how the device actually generates sound or how it differentiates between distinctly colored objects in terms of distance, intensity and saturation.[16] Another crucial problem is bone conduction. Harbisson hears colors as they resonate intracranially. Surely he hears not just colors, but the "grain" of the color, not to mention the grain of the skull, to modify Roland Barthes' famous phrase about the embodied materiality that produces and conveys (vocal) sound. But the fine-grained, bone-conducted colors can never coincide with the colors Harbisson's audiences hear at his performances. The latter are sine tones generated by his laptop and played over a PA system, and even if the laptop is connected to the eyeborg, the colors do not resound with the same granularity as the colors the artist himself hears. In so tightly closing the essentialist circuit between object, color, sound, and sensation, Harbisson effectively closes the audience *out* of the process. The artificial sense and the revealed reality are Harbisson's alone, and his audience is, by the very nature of the set-up, always on the outside, taking his word for it.

Part III. Transduction and Indeterminacy

Harbisson, consciously or not, avoids dwelling on the technical specifications that would interrupt the smooth operation of his device, and displace his cyborg body from its privileged role as the ultimate transducer of energies, translating without remainder from color to sound, architecture to music, speech to image, and even outer space to earth. Yet transduction, as I argue throughout these chapters, is a far more complex process than Harbisson is willing to admit. Transduction problematizes the energies and instruments that it simultaneously mediates, showing that it is impossible to think about sound without implicating the bodies, discourses, and institutions that structure, disrupt and transform the seemingly mechanical transposition of acoustic vibration into electrical energy.

Adrian Mackenzie has devoted an entire book to the indeterminacies and contingencies that are built into the technical and philosophical concept of transduction (2002). He merges deconstructive approaches to the problem of technology with the work of Gilbert Simondon to develop a concept of transduction as a method for thinking (about) technical systems; in particular, how they modulate differences and persist through time by holding open discrete margins of indeterminacy. Building from the technical notion of a transducer, Mackenzie understands transduction as a technical and conceptual process that operates across networks, infrastructures, and all technical systems broadly considered:

> In electrical and electronic engineering, transducers convert one form of energy into another. A microphone transduces speech into electrical currents. For the process of transduction to occur, there must be some disparity, discontinuity, or mismatch within a domain; two different forms or potentials whose disparity can be modulated. Transduction is a process whereby a disparity or difference is topologically or temporally restructured across some interface.
>
> (25)

Transducers modulate or restructure differences, but they do not equate or eliminate them. Differences remain, albeit in the form of a technical ensemble that reticulates them in space and time. For Mackenzie, following Simondon, cybernetic devices reiterate on the level of information the transductive function of simpler devices like microphones and speakers. "Devices transduce information, understood as a margin of unpredictability in a sequence of signals, into determined forms. Any device that retains a margin of indeterminacy can transduce information" (26). This retention of a margin of indeterminacy—or, in non-informational terms, a kind of receptive open-endedness that persists within operative processes of communication—describes the way in which devices, machines and technical systems maintain a kind of discontinuity to themselves, a transductive gap or suspension that is nevertheless constitutive of their very operation: "Transduction arises from the non-simultaneity or metastability of a domain, that is, in the fact that it is not fully simultaneous or coincident with itself. Boundaries, singularities, and differences underlie transductions" (17).

Among the numerous specific examples in the book is a lengthy discussion of Stelarc's *Ping Body*, a performance in which the artist wires his body to a complex array of internet connections, biofeedback devices, audiovisual equipment, and muscle-activating electronic devices. Stelarc essentially generates a kind involuntary dance performance in which variable internet speed diagnostics (or "pings") cause certain muscles in his arms, neck, and upper body to contract, and these movements in turn trigger certain kinds of pre-programmed sound effects. Biofeedback devices monitoring heart rate and blood pressure further inflect the sound effects as well as the movements. Several cameras—some mounted on the artist's body, others on constantly moving robotic arms—play back the performance and join the general cacophony of embodied audiovisual-biofeedback and internet data. Stelarc's body is undeniably *there* as the transducer that makes the whole performance happen. Yet his body is also buried, fragmented, dominated by the apparatus, sounds and images that seem to overwhelm him more than he controls them. Mackenzie emphasizes that Stelarc, on the one hand, meticulously diagrams the set-up of the performance (indeed, Stelarc makes a habit of foregrounding the technical and physiological set-ups as constitutive of, rather than supplementary to, his performances), while, on the other hand, having at best a vague, probabilistic sense of how the piece will actually unfold and what its aesthetic effects will be. This is crucial for performing the cyborg body as a transducer, that is, for the instantiation of a system that plays on the convertibility of the determinate and indeterminate, that exteriorizes or temporalizes itself in excess of, yet congruent to its very operation. That Stelarc uses diagnostic tools meant to calculate delays or lags in internet transmission speeds offers a kind of formal contrast to a piece that approaches an instantaneous fusing of the body and data, suspending yet animating Stelarc's body in an incommensurable machine world of voltage differences. The "meaning" or effects of the piece are difficult to articulate, yet this very open-endedness accomplishes the conceptual aim of staging the body as a transducer, that is, a disjunctive figure of instability and indeterminacy.

Transduction plays an important role in phenomenological approaches to the body and the senses, suggesting that the world of color/sound equivalences that Harbisson creates does not grasp the depth of the transductive processes at the heart of body and world. Mark Hansen, in his extensive engagement with the work of Merleau-Ponty as a means of conceptualizing the body and aesthetics in digital art, uses the concept of transduction to describe the body's openness to the world through simultaneous and mutually constitutive processes of separation and joining; of activation across boundaries; and of the body's coupling of primordial tactility (a kind of ur-sense) with originary technicity.[17] Though I flag my reservations about Merleau-Ponty's phenomenology in Chapter 1, I want to note here that at the heart of Hansen's argument (via Merleau-Ponty) is the notion of *écart*—a gap, separation or hinge that structures sensation as the mutual becoming of body and world through transduction. The gap discloses "the ontological principle of bodily life as fundamentally noncoincident with itself, as sundered by its essential openness to the world" (Hansen 2006: 76). He elaborates:

> The *écart* in this sense is a marker of reversibility as a necessary condition of phenomenal experience, of sensation per se. It is nothing other than this fundamental dehiscence that explains the body's need for the world (and also the world's need for the body, being's need for manifestation or phenomenalization). As such, *écart* also prevents the body from achieving pure immanence; it is that which renders it an essentially incomplete "unity;" a process of individuation that will never be fully accomplished.
>
> (72)

The conceptual vocabularies between Hansen and Mackenzie do not exactly overlap, yet the incomplete unity of the body that Hansen describes has clear affinities to Mackenzie's notion of the constitutive incompletion of technical ensembles (of which the living human body is certainly an example). Bringing these two discourses together, that is, *transducing* them, gives us something like a conceptual cyborg, a body as an interface with/in technical ensembles, which are themselves interfaced with a world (if not an earth, in their formulations) across the thresholds of sensation and information.

In my analysis of Harbisson's work, we can see that the profundity and indeterminacy that thinkers like Hansen and Mackenzie bring to transduction are not sufficiently mobilized in the cyborg art. Beyond its promotional aspects for a brand of neo-cyborgism, Harbisson's art is interested mainly in closing gaps, finding perfect equivalences, creating synesthetic senses with paradoxically reductive, tautological effects. If anything, Harbisson's sonochromatism reminds us of the fundamental incommensurabilities that structure our experience of being in the world. His scale, after all, is based on an arbitrary conjoining of tones to transpose values that not even Newton could resolve satisfactorily. Rather than play on this problem and examine the ways in which sound and music both shape and elude quantification,[18] he brands himself the creator and owner of an artificial sense. Harbisson foregoes the transductive listening through the soundscape in favor of the "private property" of customized listening.[19] His transduction of color to sound relies on a simple, even tautological formula that reduces rather than resounds the material, technical and aesthetic indeterminacies of the transductive process.

Part IV. Soundscapes without Sound

To put Harbisson in the context of another artist of roughly the same generation whose creative practice both thematizes and defamiliarizes the sociality of the senses, I want to conclude by considering the work of American Deaf artist Christine Sun Kim, who creates visual works, performance pieces, and installations that draw attention to sound as a social and environmental field. Her work foregrounds the intersensoriality of aesthetic experience, while drawing into sharp focus the audist prejudices that inform normative conceptions of voice, sound, and listening. As is the case with Harbisson, Kim is an artist working with a perceptual element of which, it would seem, she has no direct experience. Yet, as I show, Kim creates a body of work which reconfigures an apparent gap in her sensory experience as a constitutive element in her creative process, challenging norms and conventions around embodiment and communication, and exploring the possibility of inclusive, interactive deaf and hearing environments.

The 2013 piece "Courtier as Courier: Voiceless Lecture" by Kim serves as an introduction to the complex remediation of speech, textuality, gesture, and space that characterizes her work with sound. I myself have not attended a performance of the lecture, and I base my analysis on a short performance video edited by Kim herself (recorded in March 2013 at Drifter Projects in New York City) and subsequently posted on the web.[20] But, as I will argue, my own partial, belated, and subjectively reconstructed encounter with the performance is very much of a piece with the communicative problems that the work performs. In the video of "Courtier as Courier," Kim delivers a lecture based on a curious anecdote recounted in Baldassare Castiglione's *Book of the Courtier* (1528), in which an Italian merchant abroad in Poland undertakes to buy a quantity of sable pelts from Russian trappers. Because of an ongoing military conflict, the two parties are forced to address each other from the opposite banks of the frozen Dnieper river, a political boundary which everyone involved is reluctant to cross. According to the story, the Russians shout out their price, but the air is so cold that their words freeze halfway across the river before they reach the ears of the Italian, at which point his Polish companions venture out to the middle of the frozen river to build a fire and thaw out the frozen words. After an hour (by which time the Russians had quit the scene), the air above the river grows warm enough so that "the words (which had remained frozen for the space of an hour) in due course beg[i]n to melt and to fall in a murmur, like snow from the mountains in May" (a quote from Castiglione's text [p. 133] which Kim cites verbatim in her lecture). Having finally heard the message, the Italian merchant decides to back out of the deal because the asking price is too high (and also, one has to assume, because the now-absent Russians had already made the decision for him).

Kim delivers a lecture about this fanciful story through the interlinking use of digital tablets, which have been pre-loaded to display—one word at a time—the sentences that compose her lecture. There are clusters of tablets and cell phones at different "stations" around the room, some stacked on narrow shelves mounted to the walls, others just resting on the floor, waiting to be picked up and swiped by the performer at the opportune moment. Kim moves back and forth across the room (with a small audience, on their feet, following her every step of the way), searching for the tablet that will display the right word to complete a given sentence. Sometimes she swipes the "wrong" screen, skipping several words ahead in the syntax, and then has to swipe across two or three other devices to catch up. Occasionally, a needed word fails to turn up on the pre-loaded tablets, and an assistant quickly finds her a fresh device into which she can program the missing word. As she scrolls across tablets, the sentences unfold in a nonlinear manner, with repetitions, interruptions, and ellipses, and become erratically spatialized: the audience reads not just from left to right, and right to left, but from one screen to another, and indeed from one side of the room to the other. At times Kim even swipes across two devices simultaneously so that the audience reads a single sentence unfolding from two different internal points in its syntax. To depict the moment when the shouted-out fur prices freeze in mid-air, Kim programs a series of dollar signs to flash and flicker on a cluster of devices which she arranges in a corner.

In this lecture, the linguistic unit of the sentence becomes a complexly mediated choreography of screens, swipes, and unpredictably spatialized syntax. Kim's lecturing style is not the one-way dissemination of a prepared text, but an embodied performance, rife with back-tracking and improvisation, which communicatively activates every corner of the room. There are two significant metaphorical transformations at play here: on the one hand, the profusion of tablets and movements that comprises the lecture fills up the entire space and, in a way, *reverberates* throughout it, bouncing off the walls like the sound of a spoken voice. On the other hand, the audience is reading the lecture word by word, that is, gesture by gesture or swipe by swipe, and in a sense, reading English sentences presented as a kind of sign language. Kim's spatialized presentation of text takes on the reverberant power of sound, and the reading powers of the audience perform the *listening* reception of speech via gestural-textual signs. Eliza Tan has remarked that, in the context of sound's complex relationality in Kim's work, it is "[l]istening [that] serves as an indispensable fulcrum for making accessible the interval between Deaf and hearing worlds;" more than just a response to sound, listening "encompasses the meaningful conveyance of nonverbal and non-cochlear experiences mediated by other sensory facets" and "involves the whole body as a source of understanding" that is "at once personal, performative, and socially engaged" (Tan 2020: 247). Transforming text into speech and reading into listening, Kim creates an inclusive interaction in which unspoken speech is listened to as sign language, without any words being uttered aloud by the lecturer and without any prerequisite knowledge of American Sign Language (ASL) on the part of the audience.

The tablets, I want to stress, do *not* have a supplementary captioning function which preserves the integrity of a spoken voice directing the lecture. Instead, the tablets distribute voice across the gestures of Kim's embodied performance, across the space of the room, and across the audience's attention, which assembles the often disjunctive sequence of words/signs into their received/perceived version of the lecture. Rather than prosthetically standing in for an absent voice (and indeed, for the oralist norm of vocalized speech at the core of communication), the screens reimagine communication as an interactive process in which two parties meet each other halfway (as if across a frozen river), not so much to transmit a fixed meaning as to attend to the transfer and transformation of meaning's contingent possibilities. After all, the prices in Castiglione's story do not just circulate as monetary values: they physically transform into (frozen) air, then metaphorically transform into a "murmur" of snowfall, a beautiful sound/image that is arguably worth more than the overpriced pelts. Like that mixed metaphor, the communicative process enacted in Kim's voiceless lecture uses technology to document the gaps, delays, and spontaneous transformations of a reimagined communication process that productively operates across sensory and signifying modalities.

If the lecture effectively redistributes voice across such an alternate model of communication, why might Kim still insist on the subtitle "voiceless lecture"? Here, I suggest, is where the artist's Deaf politics come subtly into play. The lecture is "voiceless" because Kim does not communicate in her native language, ASL, and does not work with an ASL interpreter. As she says in her TED talk titled, "The Enchanting Music of Sign Language" (a "voiced" lecture which Kim delivers in ASL with interpreter Denise Kahler-Braaten):

> There's a massive culture around spoken language. And just because I don't use my literal voice to communicate, in society's eyes it's as if I don't have a voice at all. So I need to work with individuals who can support me as an equal and become my voice. […] I work with many different ASL interpreters. And their

> voice becomes my voice and identity. They help me to be heard. And their voices hold value and currency. Ironically, by borrowing out their voices, I'm able to maintain a temporary form of currency, kind of like taking out a loan with a very high interest rate. If I didn't continue this practice, I feel that I could just fade off into oblivion and not maintain any form of social currency.
>
> Kim 2015

For a native speaker of ASL addressing a hearing audience, voice is already an inherently distributed concept. Her voice doesn't "belong" to her in the same way a hearing person might imagine ownership of their voice, but rather emerges out of a collaborative process that grants her limited and high-cost access to the communicative world of sound. However, in "Courtier as Courier," Kim sets herself the difficult task of communicating without recourse to the language she is fluent in, and without the collaboration of an interpreter who would serve as courier for, and guarantee the currency of, her speech. She suspends her voice and proposes to the audience an alternate, temporary form of social currency.

The lecture also allows the audience to experience a similar constraint. In the video, Kim concludes her lecture by displaying two tablets that read "thank" and "you" respectively, and, with a smile, she claps her hands once or twice and then gives the ASL sign for applause (the only instance of signing in the video). In the instant the lecture ends, the communicative bond established through Kim's presentation is broken, and unless the audience members happen to know how to sign, they cannot personally express their appreciation of the performance, ask the artist a follow-up question, etc. Perhaps in a nod to Castiglione's text, they cannot be *courteous* audience members, but instead must stand in silence, trying—and probably failing—to imagine the right gesture to express their thanks in an unfamiliar language. In this regard, the conclusion of the lecture makes the audience experience a voicelessness, too, in their inability to sign, and in the absence of an ASL interpreter. The lecture not only establishes a certain equalizing communicative interaction that approaches speech (from the perspective of the Deaf artist) and signing (for the hearing audience), it also allows both artist and audience to share the experience of voicelessness that frames deaf interactions in a hearing world.

While many of Kim's works are, like "Courtier as Courier," highly mediated, making use of audio technologies, visual projections, and touch screens to choreograph communication as a multimodal process, I want to turn now to the more lo-tech series titled "The Sound of" (2016), which consists of charcoal drawings on paper. The two basic visual elements in the series are "f"s and "p"s (corresponding to the musical dynamics *forte* and *piano*), which Kim uses as a kind of handwritten binary code to represent loudness not necessarily as sound, but as affective intensity. In "The Sound of Laziness," clusters of seemingly randomly conjoined "f"s and "p"s—sometimes just two or three letters, at times as many as ten—are arranged as rows of text on the paper (Figure 4.1). Each cluster of letters is linked by an equal sign (e.g., pppf = pp = p = ppp = pp, etc.). As you "read" the drawing left to right and top to bottom, the frequency of "p"s comes to predominate, visualizing laziness as an equalization of quantitative and qualitative differences, and a steady diminuendo of affective energy. Likewise, in "The Sound of Apathy," we encounter the same visual figure over and over again (fp = fp = fp), and as we "read" through the lines of text, the *piano* response grows longer and longer, so that the last several elements of the equation read "fppp = fpppp = fppppp = fppppp = fp" (Figure 4.2). The same stimulus (*f*) triggers the same response (*p*), and no matter how much of that response there is (i.e., how many "p"s accumulate in each individual element of the set), they equate to the same value in the end. The piece depicts the diminished affective responsiveness of apathy through the visual monotony and mathematical reduction of sonic symbols.

"The Sound of ~~Resignation~~ Being Resigned" invites the audience to consider another negative emotional state (resignation) from an implicitly deaf perspective (Figure 4.3). The drawing consists only of "p"s (seven rows of eleven "p"s each), with a "less than" sign interposed ("p < p < p < p [...]"). Mathematically, the first "p" represents the quietest stage of the sequence, which traces an ever-intensifying sonic-affective state that represents itself monotonously (and paradoxically) as a crescendo of *pianos*. The crossing-out of "resignation" in favor of "being resigned" in the title creates a verbal pun across textual and sign language: being "resigned" is both the cause and effect of "re-signing" (i.e., making the same sign over and over again), with dwindling effectiveness. The drawing invites an audience who may have no knowledge of ASL to think across words, signs, gestures, and repetition to conjoin the act of re-signing (because of not being understood, not being heard) to an emotional state of being resigned.

Kim's 2018 series "Degrees of Deaf Rage" (charcoal on paper) explicitly thematizes the artist's lived experience as a Deaf person in an audist society. The series visualizes Deaf rage via angles drawn in free hand (as if with a protractor in a high school geometry class). Each angle is labeled according by its size (from "acute" to "full on"), and the degrees of the angle are colored in thickly with

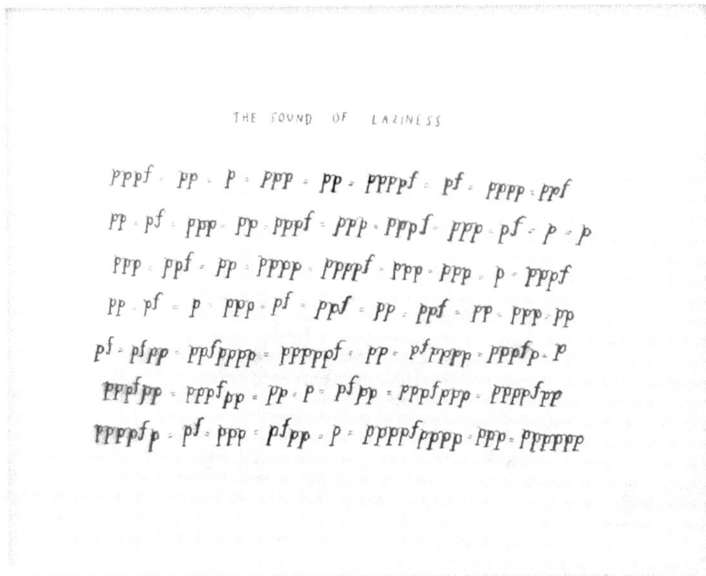

Figure 4.1 Kim, *The Sound of Laziness*, 2016. Charcoal on paper, 19.5 x 25.5 inches (49.5 x 65 cm). Courtesy of the Artist and François Ghebaly Gallery, Los Angeles. Photo: Kell Yang Sammataro.

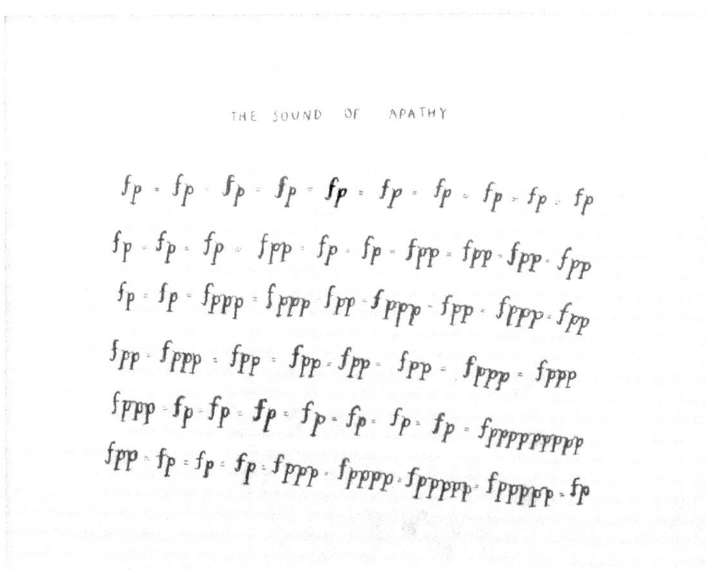

Figure 4.2 Kim, *The Sound of Apathy*, 2016. Charcoal on paper, 19.5 x 25.5 inches (49.5 x 65 cm). Courtesy of the Artist and François Ghebaly Gallery, Los Angeles. Photo: Kell Yang Sammataro.

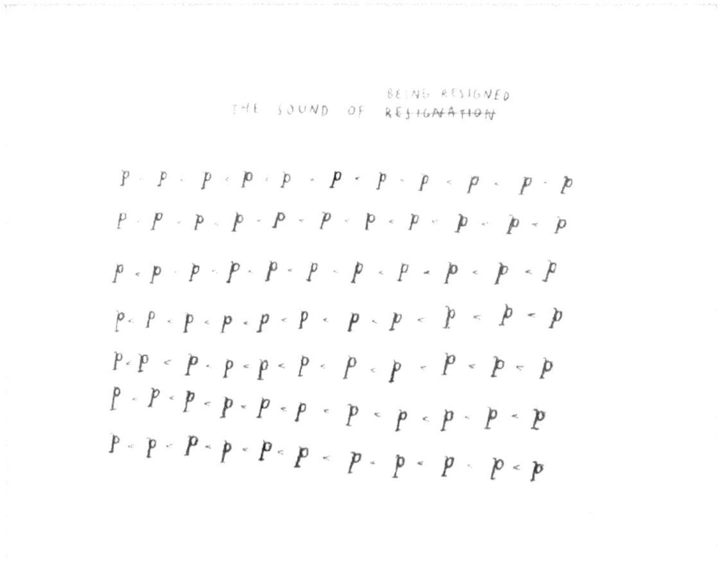

Figure 4.3 Kim, *The Sound of Being Resigned*, 2016. Charcoal on paper, 19.5 x 25.5 inches (49.5 x 65 cm). Courtesy of the Artist and François Ghebaly Gallery, Los Angeles. Photo: Kell Yang Sammataro.

charcoal. Each angle is accompanied by a brief handwritten explanation of the situation or experience that triggers that particular intensity of rage. For example, in "Degrees of Deaf Rage within Educational Settings," acute rage is explained by the caption "CANNOT ENROLL IN THE CLASSES WE WANT BECAUSE ONLY THE CLASSES THAT ARE MOST POPULAR AMONG DEAF STUDENTS GET INTERPRETERS" (Figure 4.4). At the other end of the spectrum is the "full on" or full-circle, 360-degree rage summed up by the caption "THE MILAN CONFERENCE OF 1880," which refers to the first international conference of deaf educators, and which led to the banning of sign language in all deaf educational settings in Europe and North America, paradoxically installing oralism at the heart of deaf educational and institutional life at the turn of the century. This drawing creates a scale between the minor or "acute" frustration of limited access to interpreters across the university curriculum to the historical "full-on" rage at institutionalized prejudice that was a generational set-back to deaf learners and educators, as well as a profound curtailment of international sign languages. Non-deaf audiences and people with little or no connection to the Deaf community can view these drawings and very quickly

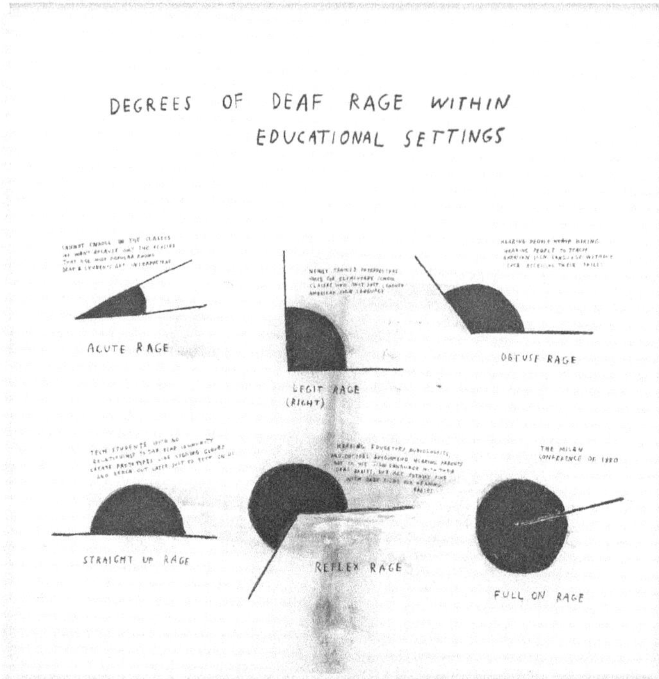

Figure 4.4 Kim, *Degrees of Deaf Rage within Educational Settings*, 2018. Charcoal and oil pastel on paper, 49.25 x 49.25 inches (125 x 125 cm). Courtesy of the Artist, François Ghebaly Gallery, Los Angeles, CA, and White Space, Beijing. Photo: Yang Hao.

acquire some necessary tools and information to empathize with deaf people. For those unaware of the historical significance of the Milan Conference, the drawings also educate non-deaf audiences and provide historical context to lived experiences of inaccessibility.

In "The Sound of" series, Kim plays effectively with conventional markers of sound (musical dynamics) to evoke sound by visualizing it in pictorial form; and in "Degrees of Deaf Rage," she shows how the implicit sonicity of the visual field is inseparable from how audism inflects that field socially and historically. In both series, the freehand nature of all drawings—the irregular spacing of the rows of the text and arrangements of the angles, the smudges caused by the writing hand moving across the charcoal-inscribed surface, the idiosyncratic shaping of each letter and symbol—add a textural vibrancy to the "content" that the drawings display. In this sense, the drawings are recordings of a hand's movement, bearing witness to the ideas and affects that animated that movement. The drawings thus reproduce a visual rhythm and timbre that situate their pictorial aspect in a sonic field. Michael Davidson sums this up succinctly: "[Kim's] drawing and sound installations make visible and tactile sounds that as a deaf person from birth she can't hear but that, at the same time, enable her listeners to hear through a deaf optic" (2021: 219). Indeed, it is difficult to characterize these drawings as either visualizations or sonifications of emotional states since they depict how an affective intensity flows through and is articulated by sonic, textual/textural, gestural, and visual representations. Kim blurs the line between a visual and an audio work, drawing attention to the ways in which the visual field is inscribed in sound through inferred, symbolic, and textural modalities, and likewise how the experience of sound is inextricable from the gestural, textual, and spoken language we use to represent it. In Kim's drawings, the sensory indeterminacy of where the visual ends and the sonic begins is thus folded into larger questions of how affect is sensed and represented; how an isolated emotional moment scales into a mood, an environment, or a historical context; and where individual experience is forced to interact with institutional prejudice and inaccessibility.

Seen in this context, we can attend to Kim's later charcoal drawing *The Sound of Temperature Rising* (2019) as a visualization of climate change that works in the same sonic-pictorial language of these affect-based pieces (Figure 4.5). We see four curved lines that recall musical slurs connecting four half notes at the lower center of the paper to sets of four quarter notes floating on the page's upper right side. The piece depicts an upward movement from a quieter, more sonically spacious state to a louder, more crowded state near the limits of what will actually fit on the page. As a kind of allegory of climate crisis, the work makes one kind of political point about how globally rising temperatures are plain to see. But the lines are significant because they may not only represent slurs, they may also indicate the artist's idiosyncratic representation of a musical staff. In "Enchanting Music," Kim explains that her pictorial rendering of the Western music staff (which contains only four lines instead of the standard five) is derived from her version of the ASL sign for "music staff," which involves dragging four extended fingers horizontally across the chest. For example, she uses a four-lined musical staff in her drawing "TBD TBC TBA" (2015) in which three empty music staffs (labeled respectively according to the drawing's title)

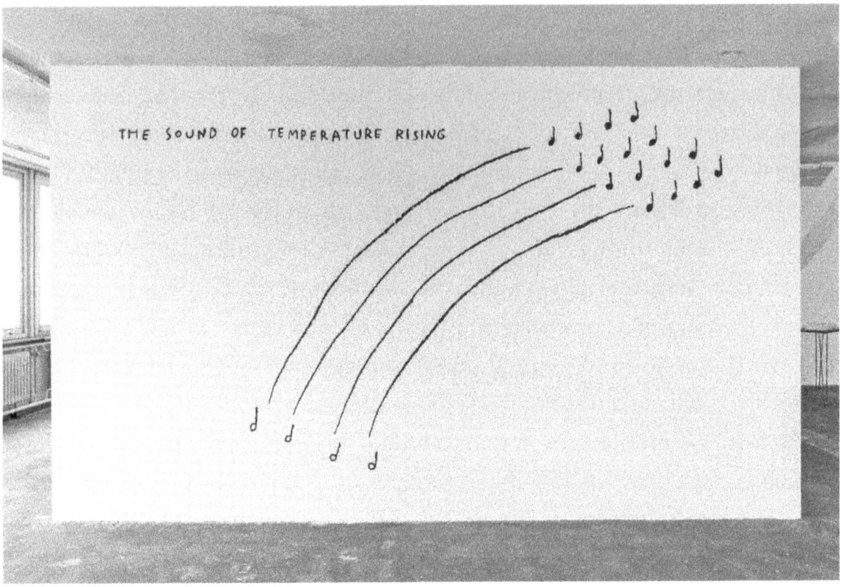

Figure 4.5 Kim, *The Sound of Temperature Rising*, 2019. Acrylic on wall adapted from 2016 charcoal on paper drawing, Edition of 3, 1AP, 115.75 x 196.75 inches (294 x 499.74 cm). Installation view, PS120, 2019, Berlin, Germany. Courtesy of the Artist and François Ghebaly Gallery, Los Angeles. Photo: Stefan Korte.

represent the indeterminate expectation of sound from a deaf perspective, according to which sound is a social event that is always determined, continued, and announced by hearing people (Figure 4.6).

Looking back to "Temperature Rising," the upward curvature of the notes *and* of the staff represents a double-entendre on the idea of scale as both a framework of measurement and a musical scale. In global warming, the drawing seems to suggest, it is not just the temperatures that rise (like notes on an ascending scale); it is the entire scale itself of ecological sustainability that is altered and radically called into question. Experienced alongside the other "Sound of" and "Deaf Rage" works, it is clear that the environmental scale of "Temperature Rising" is folded into the same affective, experiential field of individually embodied emotions. The two phenomena speak the same sonic-pictorial language, or are made to communicate aesthetically, through the drawings which transduce affect and environment into a continuous pictorial soundscape. Thus hearing through a deaf optic, we glimpse an imaginative soundscape without sound that conjoins planetary and human scales in the same geosonic language, a

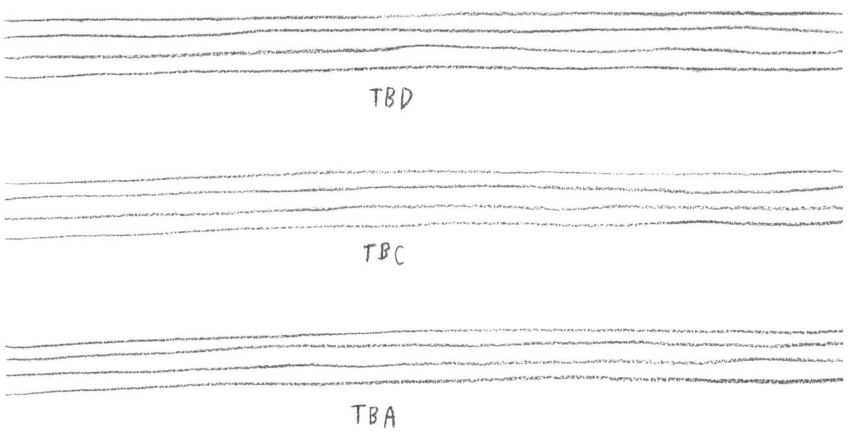

Figure 4.6 Kim, *TBD, TBC, TBA*, 2015. Charcoal on paper, 11 x 15 inches (28 x 38 cm). Courtesy the Artist, White Space, Beijing, and François Ghebaly, Los Angeles. Photo: Yang Hao.

transductive medial act all the more astounding for its technical simplicity (charcoal on paper). The indeterminacy and intersensoriality which Kim holds open across channels of communication and unheard sound becomes the aesthetic framework in which the planetary resounds as an affective soundscape for deaf and hearing listeners.

5

Sound Asleep

Sleeping, Listening, and the Politics of Nonconscious Experience

In the previous chapter, I concluded with an opposition between two types of artists, each working with sound to aesthetically configure a missing element of their respective sensory fields. Christine Sun Kim's approach is to distribute sound across visual, tactile, gestural, and textual modalities, effectively re-ciphering sound into a pictorial language that communicates with deaf and hearing audiences. Keenly aware of the interstitial spaces where the affective power of sound activates adjacent sensory and semiotic registers, Kim explores sensation and representation as fundamentally intersensorial and intersubjective processes. Neil Harbisson, by contrast, seems to reinforce the separation of the senses into distinct channels of perception, foregrounding an aesthetic of informatic transposability in which senses are owned and enhanced, rather than idiosyncratically experienced and socially shared. As stated on the Cyborg Foundation's website, the cyborg aesthetic locates "the artwork, the audience, and the museum" in the self-same body of the artist, using sound to seal the gap in the soundscape opened up by the cipher of color. Kim, by contrast, sounds through the sonic gap in the sensorium to create dialogical and critical soundscapes that invite listeners of all hearing abilities to engage and reflect on the sensation, mediation, and power relations of sound in social and historical space.

In setting up this opposition, I do not mean to favor one artist's creative practice over the other, but rather I demonstrate that each artist's work sustains a particular kind of geosonic attention. Kim's deceptively simple drawing of rising temperatures raises, when seen in the context of her work on communication between deaf and hearing worlds, complex questions about the clashing scales of sound, affect, and environment in an era of climate crisis. It is, in my view, as compelling a depiction of the sound of climate change as many works of ecological sound art that use "actual" geophysical sounds or sonified data. In Harbisson's case, when I listen geosonically to the colorized sine tones in his artworks, what I hear is not so much color as a certain sonic-ideological premise: namely, that digitized sound (via increasingly mobile, wearable, 24/7 delivery technologies) can and should smooth out of the inconsistencies and indeterminacies of embodied experience, and that sound's materiality can and should be appropriated as a neoliberal technology of the self. Despite boldly reimagining the body as a technical interface and transducer of earthly energies, Harbisson actually sonifies a normative framework in which dominant subjects use (audio) technology to control themselves and their environments.

In this chapter, I expand on the body's affective entanglements in earth-scale systems, energies, and ideologies, critically analyzing sound's power to reveal or conceal those entanglements. Starting from the geosonic premise that the experience of being in a body is inextricable from being on a planet, I posit sleep as a crucial interface between embodiment and what I call "enearthment," the earthly, planetary embeddedness of bodily experience. I theorize sleep as a geosonic process in which planetary cycles of light and dark interact with the human brain, crystallizing sleep as the by-product or processing artefact of informational and neurochemical processes. Sleep, thus defined, recasts the brain not as a center of rational cognition and consciousness, but rather as a relay of nonconscious (but not necessarily noncognitive) embodied processes. Sleep is nevertheless part of our lived, daily/nightly experience, and as such also represents a liminal or paradoxical nonexperience: in sleeping, we live out an obscure, but immanent connection between body and planet though the affectivity of shared rhythms and periodicities. Privileging sound as the material and metaphorical means of articulating the embodiment and enearthment of sleep, I borrow from theories of affective listening to explore the affective planetarity that enfolds and permeates our bodies in ways analogous to the vibratory affectivity of sound. I conclude with some critical remarks on the neoliberal biopoliticization of sleep though an analysis of Max Richter's *Sleep* (2015b), an 8-hour piece of music composed for a small chamber ensemble, meant to be performed before—and listened to by—a live, sleeping audience.

Part I. Sleep on the Brain

There exists a lived brain, but [. . .] this lived brain is not necessarily conscious.

Catherine Malabou, *What Should We Do with Our Brain?* (2008: 16)

The modern scientific definition of sleep rests on a palimpsest of three historically distinct, cerebro-centric models.[1] The first is mechanistic and homeostatic: neurochemicals accumulate in the brain after too little sleep and gradually exert a force of so-called "sleep pressure" that increases over time. Sleep itself acts as a kind of release valve for this accumulated pressure, allowing the machinery of the post-sleep brain to work again at an optimum level. A second model is cybernetic, envisioning sleep as a complex biofeedback process, which links environmental cycles of day and night to a brain center that governs the physiological and neurological processes associated with sleep. A third model, derived from the metaphor of the brain as computer, approaches sleep as a kind of informational process in which the brain organizes and stores data acquired in the course of the day for future retrieval and use. In any case, the sleepy or sleeping brain is figured variously as a steam engine, thermostat, and computer, each corresponding to key technoscientific paradigms of the twentieth century. Undergirding all of this is the ultimate metric of sleep, electroencephalography (EEG), which uses electrodes placed on the scalp to record electrical differences associated with brain activity and which has given rise to the modern definition of sleep as a phased alteration of brain waves distinct from that of the conscious or waking brain. So decisive was the introduction of EEG to the sleep laboratory that one historian has remarked that "the ability to visualize sleep in terms of recorded graphical data was absolutely pivotal to the status of sleep as an object of modern scientific research and biomedical practice" (Kroker 2007). Sleep, in other words, only came into being through the real-time coupling of a scientific instrument with the brain.

In this sense, long before the development of more sophisticated techniques for monitoring and modeling brain activity in real time—like the cutting-edge functional magnetic resonance imaging (fMRI) of today—sleep had quietly inaugurated a paradigm in the neurosciences I call "brainism"—a predominant tendency to isolate and monitor specific areas of brain activity, usually in highly artificial and highly technologized laboratory environments, and then to ascribe complex mental events like consciousness, memory, volition, et cetera to those "legible" areas. Brainism implies the downplaying of language, culture, environment, and daily life in the scientific study of the brain, yet insists at the same time that its minute, laboratory-specific observations can serve as the explanatory ground for cerebral and mental events that necessarily occur out there in the world. Sleep, in a sense, paved the way for this general bracketing or reduction of the embodied and environmental dimensions of the (study of the) brain: by conjoining a specific technology (EEG) to specific brains in a real-time laboratory environment, sleep first produced the brainist epistemological conditions by which scientists could claim to know and describe complex mental and embodied processes primarily by measuring and visualizing brain activity. Sleep in neuroscience is analogous to the eardrum in audio technology (as outlined by Jonathan Sterne in *The Audible Past* [2003]) insofar as the physiology of an isolated organ grounds an epistemological paradigm for treating a host of technical and ontological questions. In short, brainism is like tympanism. Despite critiques from a range of philosophical, technical, and neuroscientific perspectives, a certain brainism still predominates in contemporary visualizations of the functional brain, in which hi-tech imaging offers an ontological short-cut to age-old debates around minds, brains, and the complexity of human thought (Rose 2016).

If there is such a thing as a lived, embodied brain—that is, a brain that cannot be technically reduced to cerebral monitoring and isolated from the body to which it is attached, not to mention from the social and geographical environment in which it is immersed—then perhaps one way to approach it is to rethink sleep. In order to do this, first I turn to Catherine Malabou's concept of neural plasticity (2008). From a critical and philosophical perspective, Catherine Malabou has argued effectively against both literary metaphors and scientific models that cast the brain in the role of an autonomous control center, that is, against the brainist tendency to regard the brain as a discrete, biologically determined and technically visualizable entity that serves as the ultimate biological and metaphysical ground of consciousness and mental life. In contrast to the perceived brainism of the cognitive and neurosciences, Malabou's concept of the brain's plasticity (which explicitly replaces a century's worth of mechanistic, homeostatic, and informational models) stresses how the brain is shaped both internally (by biological factors and neurological processes) and externally (by environment and culture). Plasticity reflects the neuronal interaction of the brain's determinate and indeterminate development, and, philosophically, the contradiction between a closed, biological form and the openness of a mind. That openness is, at the same time, dialectically grounded in the neurophysiology of brain function: in synapses (the *literal* openings that connect neurons), and in the aleatory (re)formation of neural pathways across synapses over time. For Malabou, the brain is neither a metaphor, nor an identity: it is a field of action constituted by plasticity and as such, it resists not only biological determinism, but also determinism's equivalent in socioeconomic terms: neoliberal globalism (to which I return in Part II).

Against the brainist discourses of neuro- and cognitive sciences, I would suggest that Malabou conceives of the brain like a body. Taking the above epigraph one step further: that which is lived, but not necessarily conscious is precisely the body, and the embodied brain along with it. Sleep, then, would be a privileged means of glimpsing or sounding out the lived brain because sleep bypasses consciousness altogether and reveals the embodiedness of the brain as such. Yet how can we locate and measure the sleep of the body in the absence of a privileged technological device like the EEG? The answer requires a shift in scale via the cybernetic model of sleep I mention above. If sleep is a kind of planetary networking, then the sleeping body can only be found on the rotating body of the earth, and these two bodies, from the perspective of sleep, mutually determine each other. The interdisciplinary field of body studies already grapples with the fundamental problem of how to define and even locate the body in terms of its inside and outside; its lived, subjective dimensions; its cultural and social enmeshment; its technological coupling; and its environmental embeddedness.[2] Studying the body means inquiring into the boundaries of what constitutes the body (in relation to medicine, media, technology and social life), even as those boundaries are constantly being reimagined and renegotiated. The *sleeping* body extends that shifting borderline to include the light–dark cycles of planetary rotation and forces us to conceptualize the body's interaction with and embeddedness in not only social and cultural life, technological assemblages, and physical environments, but also in the rotational motion of the earth and the rhythms it imposes, in short to rethink embodiment as enearthment.

The cybernetic model of sleep is an entrainment of internal bodily processes with planetary motion, whereby minute quantities of neurochemicals and hormones ebb and flow in sync with the rotation of the earth. The body's circadian rhythm requires constant feedback and recalibration from external light sources (above all, the light of day and the dark of night) in order to function in proper alignment with the homeostatic and informational needs for sleep. Here it is important to note the literal meaning of "circadian," namely "about a day." The in-built rhythm of numerous cellular and metabolic processes cycles only *roughly* every 24 hours, but the mechanism requires regular environmental cues to synchronize the dual processes of the sleep-wake cycle.[3] Without the input of external lighting cues, a deviation of even a few minutes +/- 24 hours on the internal homeostat can produce chronic sleep disturbances in individuals. Lockley and Foster write, "The 24-hour light–dark cycle is the most important environmental time signal, and this light information is captured exclusively by the eyes in mammals" (2012: 20). They go on to point out that the human eye contains within it a small percentage of cells—photoreceptive retinal ganglion cells or pRGCs—whose sole function is to transmit light directly to brains areas responsible for the regulation of circadian rhythms (21). In other words, about 1-3% of the human eye has nothing to do with vision at all, but is devoted exclusively to illuminating the interior of the brain with daylight.

Night is also registered by a different part of the brain, the pineal gland, which further contributes to the synchronization of the sleep–wake cycle by secreting the hormone melatonin: "Pineal melatonin is the major biochemical correlate of darkness and provides an internal representation of the environmental night-length. As night-length changes with the seasons in non-equatorial latitudes, melatonin not only encodes the daily night-length, but also the time of year (season)" (23). Although the exact role that melatonin plays in the onset and maintenance of sleep remains to be specified, its secretion has become the major "phase-marker" of the circadian clock in human sleep studies (24). The eye and brain are thus hardwired to respond to the rotation of the earth as it manifests itself in light *and* dark, and not just in the course of a single day, but over entire seasons and years. The sleeping brain is a kind of global positioning system for itself, constantly establishing and calibrating its location on the surface of the earth. Sleep, then, is our user interface between our body and planetary motion. In this sense, it is impossible to draw an absolute distinction between a (human) body at rest and a planet in constant motion as the former simply would not exist without the latter. Even if the brain is at the center of sleep, it is the earth that is at the center of the (sleeping) brain.

The concept of interface is inherently multiple: interfaces imply touch, control, access, and communication between ostensibly passive or inert devices or instruments, and active, usually human users. But if we take sleep as a kind of user interface, we have to ask who is using whom? Do humans use or instrumentalize the earth when they sleep? Or is the earth itself pushing our buttons and eliciting an anticipated response from us, that is, using or instrumentalizing us? What kind of agency or instrumentality emerges when we think about the nonconscious experience of sleep as a human/earth interface? Here I want to turn to Jean-Luc Nancy's meditation on sleep, *The Fall of Sleep* (2009), where Nancy sets himself the paradoxical task of a phenomenology of sleep, that is phenomenology of a state in which the two crucial, mutually constitutive elements of that philosophical practice—self and world—are suspended. The absence of self and world, for Nancy, amounts to a kind of paradoxical instantiation of embodiment and enearthment.

For Nancy, sleep is a mode of indistinction, a state in which the conscious self does not so much disappear altogether as merge or coincide with the body and the world. He writes that, in sleep:

> I become indistinct [. . .] I no longer properly distinguish myself from the world or from others, from my own body or from my mind, either. For I can no longer hold anything as an object, as a perception or a thought without this very thing making itself felt as being *at the same time* myself and something other than myself. A simultaneity of what is one's own and not one's own occurs as this distinction falls away.
>
> (7)

Sleep is thus, on the one hand, a phenomenological void, an evacuation from the self of all the perceptions, objects and, indeed, the world in which it embeds itself, a state in which there are no phenomena to be phenomenalized. Nancy even writes explicitly that "[t]here is no phenomenology of sleep" (24). Yet on the other hand, for Nancy the sleeping body is the very (non)manifestation of that *thing* at the root of all phenomena, a real instance of the Kantian thing-in-itself. He calls the sleeping body, "The thing isolated from all manifestation, from all phenomenality, the sleeping thing at rest, sheltered from knowledge, techniques, and arts of all kinds, exempt from judgments and prospects. The thing not measured, not measurable, thing concentrated in its indeterminate and non-appearing thingness" (14). The sleeping body thus somehow appears precisely in its non-appearance, *does* something precisely because it *manifests* nothing, and unfolds itself into and across the world precisely because it only burrows into itself, into its thingness.

But the sleeping body, even if it is a kind of thing-in-itself, is not isolated or cut off from other things. Sleep joins bodies in a kind of indistinct distinction and validates them according to a principle of general equivalence, of equality. Nancy writes: "Sleep itself knows only equality, the measure common to all, which allows no differences or disparities" (17). So the disappearance of the self in sleep reveals the equality of all bodies and—or with—the world, or the world's non-phenomenological counterpart, the earth. Here Nancy writes: "For it is in effect the great equal sleep of the whole Earth that those who sleep together share. In their 'together' is refracted the entirety of all sleepers: animals, plants, rivers, seas, sands, stars, set in their crystalline spheres of ether, and ether itself, which has fallen asleep" (20). Nancy's book contains many protracted, enumerative passages like this one that attempt to span the indistinction or equivalence of sleep across minute, inorganic particles; terrestrial life; and celestial phenomena on a cosmic scale. Unwieldy as they are to cite, let alone conceptualize, such passages at least suggest how sleep opens up an immense continuum, a kind of thingly togetherness or concrescence that comprises animal, vegetal and celestial bodies, both organic and inorganic.[4] This equality of sleep is not experienced, but it is shared; it is not sensed, but it is fulfilled. And it posits, in an obscure and silent way, some kind of possibility of expanded being in or on a phenomenological nowhere called the earth.

There is much to critique in Nancy's phenomenological lullaby. After all, sleep requires time, money, and shelter, and thus serves to amplify, rather than equalize the impact of socioeconomic factors on our embodied experience (Williams 2011). In the work of epidemiologists like Chandra Jackson and Carmela Alcántara who conduct population-scale studies on sleep and health, sleep in fact emerges precisely as an index of socioeconomic inequality as it intersects with gender and race (Jackson et al. 2018 and 2020; Alcántara et al. 2016 and 2017). There is moreover no single, evolutionary ur-sleep, but rather only the overlaying of internal and external rhythms that vary across organisms, seasons, and environments and depend on the availability of resources, shelter, etc. (Reiss 2017; Wolf-Meyer 2012). In short, sleep's "indistinction" is political through and through, and re-inscribes on our bodies all the differences, inequalities, and contingencies that it may suspend from our consciousness. Yet what I want to emphasize in Nancy's account is that sleep's embodiment is also a kind of enearthment. As much as sleep suspends concepts of self, place, and world, it also embeds them on the earth, simultaneously dispersing and burrowing the self into the body and into the earth, which is itself but another body among countless, but not equivalent others. Yet if we tone down the metaphorical and speculative register of Nancy's writing and transduce the infrastructure beneath its immersive rhetoric, we find that we are not too far afield from the cybernetic model of sleep, which diagrams sleep as a network linking bodies, brains, and the planet. If Nancy seems to suggest that when we sleep, we all hold a body and the earth in common, then the cybernetics of sleep suggests, along similar lines, that we sleep because we all carry a piece of the planet in our very own brains. A further affinity between these two theories of sleep is rhythm: the circadian rhythm that entrains body, brain and planet could be regarded as a technical explanation of the rhythm that Nancy identifies in the "rocking" of the body during sleep. He writes that "sleep is also a rhythm, a rocking back and forth, the body rocked to the rhythm of its heart and lungs" (2009: 30–1), and he continues in another rhetorically and conceptually expansive passage:

> Rocking is a matter of high and low and of right and left, of the great symmetries, asymmetries, and alternations that govern crystals, tides, seasons, the cycles of planets and their satellites, exchanges of oxygen and carbon dioxide, captures and releases, assimilations and evacuations, nervous systems, attractions and repulsions between metals, between flora and fauna, between sexes, between stellar masses, black holes, quarks, and infinitesimal jets of dust . . . It is a matter, to conclude or rather to begin, of the initial beat between something and nothing, between the world and the void, which also means between the world and itself. (30)

Rhythm is a kind of constitutive gap for Nancy that enables the oscillation between our sense of subjective, waking distinction and our nonsense of earthly indistinction. Our brain may not allow us to experience this rhythm, but it does nevertheless facilitate it. But "brain" here cannot be reduced to the brainist encephalon of the sleep sciences. Rather, brain is part of a geosonic ensemble of chemicals and cosmos, rhythmically implicated in the enearthment and embodiment of sleep.

Part II. Nonconscious Listening

> *How can one study something to which one does not pay attention?*
> Annahid Kassabian, *Ubiquitous Listening* (2013b: 142)

If the cybernetics and nonphenomenality of sleep constitute an affective rhythm, we can think of sleeping as akin to listening and can explore an analogy or affinity between the nonconscious experiences of sleeping and the nonconscious, *affective* experiences of listening. There are currently two related, but distinct approaches to affect and listening: one that links affect and the materiality of sound to the aesthetics and politics of the (in)audible;[5] and a second that thinks through affect, sound and space within a framework of urban geography.[6] What both approaches have in common is the reconceptualization of listening as distinct from conscious attention, yet nevertheless constitutive of how we (make) sense (of) sound, bodies and the world around us. Steve Goodman writes, "Before the activation of causal or semantic, that is, cognitive listening, the sonic is a phenomenon of contact and displays, though an array of autonomic responses, a whole spectrum of affective powers" (2012: 10). On a similar note, Annahid Kassabian writes that "all listening is importantly physiological, and [. . .] many kinds of listening take place over a wide range of degrees or kinds of consciousness and attention" (2013b: xxi–xxii). Indeed, for Kassabian the attentive listening subject need not be especially attentive, nor even a subject in the first place: for Kassabian (explicitly, and implicitly for much work on sound and affect), the listening subject is not prior to the act of listening, but is rather a residue left behind by the flow of affect (qua sound or vibration) through a body.[7] Affect thus entails transduction within, but, more importantly, beyond and beneath the culturally and discursively constructed notions of subjectivity or self. When we listen, we also listen along with the affective, bodily substrate that both constitutes and constrains our conscious, attentive listening. In other words, we listen to consciously heard sound *and* to our nonconscious selves listening to sound at the same time.[8] Or as I phrase it in Chapter 1, we listen always again to the material and metaphorical, to physics-and-physiology-becoming-culture and vice versa.

Not only is it possible to listen to sounds within a feedback loop of nonconscious "attention" to them, it is also possible to listen to sounds that do not even exist. In a subsequent piece explicitly on the theme of "Music for Sleeping" (2013a) Kassabian discusses the use of binaural beats as a sleep aid, that is, how certain sonic sleep aids propose to "tune" the brain to sleep through beat frequencies. She cites a classic audiological study by Gerald Oster, in which listeners hear sounds of minutely varying frequencies in each ear of a set of binaural headphones. Listeners would then "hear" the interference between the two signals as infra-sonic beats, even if one signal were beneath the audible threshold, or masked in background noise. As Oster puts it, "Evidently the brain is able to detect and process the signals even though one of them is too weak to impinge on consciousness" (1973: 96). Kassabian calls such beats "a processing artefact of the brain, not an actual sound" because the brain (and not the ears) hears a beat of, say, 10Hz when exposed to one 440Hz tone in one ear, and one 430Hz one in the other (2013a). There is no sound with a frequency of 10Hz emanating from any of the audio equipment in the scenario, yet the brain essentially provides that vibration, or presupposes it, and thus "hears" it. The seeming paradox that the brain hears sounds that are not there, or that a listener can listen to the neurophysiological by-products of the very process of listening is, in a way, a kind of proof of the affectivity of listening: we are affected by sounds (as vibrations) that we do not, indeed *cannot* hear, just as much as we can listen to sounds that do not exist. Sound constantly converts itself from the material to the imaginary (even hallucinatory) at the frontier between conscious and nonconscious listening.

Both Goodman and Kassabian attend to the ways in which pre- and nonconscious listening (including listening to infra- and ultra-sonic frequencies) affect listeners' bodies and become part of the overall material and cultural networks in which sounds are manufactured and circulated. Conscious, attentive listening would then be based in the nonconscious affective body of the listener, a body that registers and resonates with vibrations whether or not they are consciously perceived and meaningfully articulated. The materiality of affect thus merges with a certain materiality of sound insofar as both can be rooted in an ontology of vibration, and analyzed in and through the vibrational assemblages of bodies. This expanded notion of a body (which would include any material that can conduct vibrations) would make it theoretically possible "to hear how sound circulates through any and every kind of body: from plants, animals, machines, objects and architectures, through to environments, atmospheres and the earth itself" (Gallagher 2016: 44). Sound, as Michael Gallagher describes it here, sounds very much like sleep as I describe it above. Listening to sound affectively has little to do with hearing actual sounds, that is, with the thresholds of human audibility and the conscious perception of sound waves. Rather it is an imaginative listening that recognizes (if it is possible recognize without cognition) or opens itself up to material and environmental scales well beyond the human experience, yet within which the human is inevitably caught up—and trying to catch up.

Affect, like sound, moves faster than the speed of thought, and consciousness and self are always lagging behind the complex work of affect. A similar temporal gap is of increasing interest in recent work on cognition that explores the interactions between conscious thought and what N. Katherine Hayles calls the cognitive unconscious (2017). The human cognitive unconscious refers to all the information processing carried out by the nervous system "to keep consciousness, with its slow uptake and limited processing ability, from being overwhelmed with the floods of interior and exterior information streaming into the brain every millisecond" (10). Hayles examines how cognition is distributed across a range of conscious and nonconscious processes in the human nervous systems, as well as other biological and technical systems. Hayles' turn away from the privileged domain of consciousness offers a significant anti-brainist take on neuro- and cognitive sciences, insofar as cognition is attributed not only to an extended or distributed human nervous system, but also to plants, microbes, and machines. Such de-centering of the human and de-braining of cognition would seem to resonate not only with the affect theory mentioned above (among other new materialisms)[9], but also with Malabou's notion of the plasticity of the brain. For Malabou, as I have suggested, the plastic brain is constantly de-territorializing itself, both literally (through neuronal plasticity) and metaphorically through its slippage between the models and metaphors that try to fix and define it. Here Malabou makes a rigid distinction between the transformative potential of plasticity and neoliberal flexibility, which, for her, means a neuro-biologistic model of adaptation, acceleration and resilience that equally describes brain cells and the neoliberal workforce.[10] What is at stake in Malabou's book is the perniciousness of metaphors in locking us in to patterns of thought that traverse politics and economy and the grey matter in our skulls. The consciousness of the brain's plasticity is an important philosophical and political step in challenging and reimagining what, for Malabou, would be an emancipatory politics in political and economic terms. But does the cognitive nonconscious (and, by extension, listening and sleeping) have the same emancipatory politics?

I have argued above that the nonconscious experience of sleep is akin to a kind of nonconscious affective listening, that is, to an opening-up of the body and mind to the sensation and imagination of nonconscious affects; material and planetary scales; and to an embodied and enearthed experience that conjoins all (listening and sleeping) bodies in a shared rhythm or vibration. This would mean that sleep, like listening, is a kind of cognition (albeit a nonconscious one), in which the brain is engaged in informational and cybernetic work that does not require, or is in fact hindered by, the deliberation of consciousness. Sleep's nonconscious cognition would include the processing of sleep signals from internal sleep mechanisms—as well as from the environmental cycles of light and dark—signals whose transmission is constant. As a form of cognition and information processing, human sleep would thus be constantly catching up and recalibrating itself according to the internal and external information that, in principle, exceeds it, streaming across our nervous systems 24/7, and leaving us with only 8 hours a day (optimistically) to keep up. From this perspective, sleep would cease to be a slowing down of body and mind, a rhythmical pause or caesura of conscious and cognitive life; on the contrary, sleep would be an acceleration, a bodily speed-up to adjust and adapt to a 24/7 flow of biological and planetary information. Sleep would likewise be enfolded into the anti-rhythm of the 24/7 information economy, with its imperative of constant connectivity and its promise of an asymptotically receding up-to-date-ness (Crary 2013). It is with this dichotomy—sleep as a rhythm traversing body/earth, sound/affect, matter/imagination on the one hand; sleep as perpetual deficit that reinforces the accelerated self-management of the neoliberal individual on the other—that I turn to Max Richter's *Sleep* (2015b).

Part III. The Politics of *Sleep*

> *The sleepless are on call at any hour, unresistingly ready for anything, alert and unconscious at once.*
>
> Theodor Adorno, *Minima Moralia: Reflections from Damaged Life* ([1951] 2005: 38)

I borrow the phrase "politics of sleep" from Simon J. Williams' book (2011), with the addition of italics to signal my interest in thinking about the politics of sleep via the politics of *Sleep*, a piece of music (and an immersive, durational soundscape) composed by Max Richter (2015b). In Williams' contemporary sociological account, sleep is an ideological prism that reflects two main aspects of neoliberalism: on the one hand, sleep is a "powerful and potentially troublesome or problematic reminder of the limits of rational modernity, which in turn, paradoxically, further fuels or redoubles attempt to 'discipline,' 'contain,' 'control' or 'rationalize' it" (2011: xiii). This connects sleep to already existing discourses on (the management of) worker fatigue, weariness, and depression.[11] At the same time, contemporary sleep has become "a personal and public matter of concern, [...] if not something to be colonised, commercialised or capitalised upon through a burgeoning sleep industry which now includes everything from sleep clinics to mattresses, bedding and soporific CDs for that 'perfect night's sleep'" (xiv). Sleep finds itself in double bind insofar as it is both devalued *and* monetized, an ongoing public crisis from which we all suffer while only a few profit. Any simple exhortation to sleep (even if it comes in the form of an 8-hour music performance) would have to negotiate the two sides of the ideological, instrumentalizing trap, which both disciplines and regulates sleep, while cashing in on it under the guise of promoting it. In what follows, I argue that Richter engages this neoliberal ideology of sleep obliquely, displacing the politics of sleep itself onto a problem of listening and background music.

Sleep is a fully notated musical composition for piano/keyboards, string quintet and voice. It is sold as an 8-disc recording, as well as an hour-long extract called *From* Sleep, and is, of course, available for streaming and download online. The entire piece has been performed live a number of times since its 2015 release, with shows in London, Paris, Sydney, New York City, and, in the summer of 2018, two outdoor performances at a park in Los Angeles. Instead of seats, cots and bedding are provided for each audience member. The work is a set of variations in 31 movements, cycling around a handful of themes, stretching, elongating, and echoing the same musical ideas over the hours. It is a quiet and slow piece, as one would expect, with the keyboard (played by Richter himself) plodding out austere, minimalist chord cycles that have a kind of intensity if only through their sheer slowness and repetition. The strings play long bows, avoiding high registers, creating a dim, gauzy texture around the central pulse of the keys. The solo soprano, when it does appear, hums a kind of counter-theme to the opening progression, and, at the end of the piece, vocalizes a single repeated note for minutes on end. The variations are loosely based on Richter's understanding of the neuroscience of sleep (he even consulted a famous neuroscientist, David Eagleman, as he worked on the composition) so that the timbral and rhythmic variations of the music follow the EEG-based stages of sleep. There is thus a deep narrative structure at work in the piece, even if it isn't exactly available in isolated moments of close listening.

Stylistically, there is a kind of minimalist nod to the Baroque (and one cannot help but think of the obvious link to Bach's soporific *Goldberg Variations*) and to early modern polyphonic singing, as well as elements of contemporary ambient and drone music (Richter, liner notes 2015a). In hearing the opening bars of the piece, one cannot escape the sense that there is something happening, something to listen to. And yet, the nondevelopment of the music (at least on the minute-to-minute scale) thwarts our ability to concentrate on musical content for too long. *Sleep* asks us to pay close attention to it—and its European classical styling, concert-hall trappings and *Deutsche Grammophon* branding all chime in to legitimize that attention. Yet the music also tells us to get distracted, to daydream, or just dream, period—because it also seems to suggest that the music is going nowhere in no great hurry, and the listener should follow suit

It is tempting to hear, in *Sleep*, a kind of hymn to precisely the kind of earthly co-sleeping that I've attributed to Nancy above. Richter, in his liner notes, calls the piece a "personal lullaby for a frenetic world—a manifesto for a slower pace of existence." In a promotional video on YouTube, he says he wrote the piece out of a conviction that the "night can offer us some creative refuge" from "the fast-paced digital world" (2018). Richter, like Nancy, seems genuinely concerned with sleep as something we all hold in common, as perhaps the last dimension of the human experience under capitalism where privacy and property dissolve in a shared experience of radical equality, sociality, and ecological interconnectedness. Yet this radical ecology/equality of sleep seems incompatible with the neoliberal ideology of sleep outlined above. If even the affirmation of sleep participates in an ideological matrix that seeks to control it and sell it, then how can we think through what Richter might be (however unwittingly, and with doubtlessly good intentions, both artistically and humanly) selling us?

By composing for the background in order to make an environment more conducive to a particular human activity (even if that activity is sleeping), Richter's piece shows an affinity to so-called background music, easy-listening, mood music—what can be summed in a single, if now defunct, corporate word: Muzak. Muzak is the crystallization of the military-industrial complex in musical form, instrumental music in the most pernicious sense of the term.[12] In workplaces, Muzak's famed "stimulus progression" programming was designed to use ascending patterns in rhythm and tempo to counter, minute-for-minute, the "Industrial Efficiency Curve," that is, the psychology and physiology of worker fatigue.[13] Its off-work programming (which went 24/7 as of January 1942) was designed to relax and soothe in domestic environments (and was even proven effective in slaughterhouses), as well as to create upbeat commercial spaces for boosted sales.[14]

Goodman goes further to identify a decisive change in Muzak's compositional or design strategy. In the 1980s, Muzak shifted from the "stimulus progression" model to what he calls "quantum modulation," that is, apparent changes in tempo, rhythm, and musical "color" that in fact conceal a plateau of affective intensity. Muzak, in other words, sought to manipulate workers/consumers directly at the level of affect, to manufacture a mood, rather than discipline a body. The way to overcome the bodily and mental limits of worker/consumer inertia was to bypass them altogether and directly target the more pervasive and profound level of affect. For Goodman, this amounts to a shift from a disciplinary society to a neoliberal society of control, and it manifests itself in sonic terms by the deployment of strategies of sonic branding (viral audio, hooks, and earworms) that are meant to operate below the threshold of conscious listening. Such branding is not meant to sell particular products (as "jingles" did in the past), or even to facilitate acts of consumption (as Muzak in shopping malls sought to do). Rather it seeks to "catalyze the [very] motivation to consume," to create sonic and affective associations between virtual consumers and brands that will coalesce in the not-too-distant capitalist future (Goodman 2012: 145).

What is crucial here for my purposes is that this turn signifies the beginning of the 24/7 working/listening/sleeping day. Goodman writes, "The submerged affective sensorium in which ubiquitous listening is now a subset compels the transformation of outmoded frameworks of sonic thought [e.g., conventional sonic branding and Muzak]. In an attempt to perform the necessary upgrade, an audio virology starts from the premise of a mode of audition that is 'always on'" (145). Indeed, as we have seen, affective listening is always on, even when consciousness is not, just as the sleeper is always "on" in noncognitive terms, informationally catching up to the nonstop spiralling of the planet. It is perhaps no coincidence that Goodman's chapter on "The Earworm" begins with a literary epigraph about a sleeper awoken by a sound he heard literally *in* his sleep. The earworm is heard in and by the sleeping brain not only when conscious listening is suspended, but in the absence of any actual sonic stimuli. Unlike some nocturnal ambient sound that makes its way into the texture of a dream, the earworm heard by the "always on" sleeper/listener is a processing artefact of a brain played like an instrument, or instrumentalized, by predatory sonic branders. The actual production of "music for sleeping" and other sonic sleep aids in the early twenty-first century would merely be the formalization of a colonization of sleeping that already occurred as a colonization of listening in the background music and sonic virality of the late twentieth century.[15]

The ambitious conceit of Richter's music, its formidable compositional architecture, its daunting performance time (for both musicians and audience) are not easy to dismiss; yet it is all rather uneasily congruent with a musical aesthetics and infrastructure aimed at manipulating workers and consumers, unconsciously, indeed of creating unconscious workers and consumers (as prefigured in my epigraph from Adorno). Richter's *Sleep* may well represent a strange dialectal twist in which art music fully embraces the status of background music that, in many ways, it has already attained unwittingly.[16] And the performance of *Sleep* would amount to the simulation of a shared experience whose conditions of possibility have been ideologically and practically abolished. It is entirely conceivable that *Sleep* has been marketed as a way to leverage contemporary neurocapital, or that Silicon Valley managers would urge their teams to attend a performance of *Sleep*, only to report back to the office the next morning, ready to deliver and innovate, etc. From the side of the listener, it would seem that *Sleep* would resonate in imaginative and affective spaces that have already been sounded out by capitalism, with the embodied/enearthed experience of sleep nodding off to, rather than resisting the 24/7 world.

To return to the dichotomy I invoked at the start of this analysis, sleep is a geosonic process that enacts a rhythmic connection between the human body and the planet through what I argue is the shared (non)cognitive and affective infrastructure of sleeping and listening. But that same model also positions sleep as a neoliberal speed-up, in which consciousness must optimize the 24/7 informational processing of its environment by other, nonconscious, means. From this split perspective, I suggest that *Sleep*'s promise as a soundscape of ecology and equality is undermined as much by its tacit playing to the background as by its overt performance premise. The immersive environment of the concert hall prescribes a fixed beginning and ending to the piece, and physically distributes listeners throughout a highly regulated (albeit recumbent) social space. Even if listeners consume *Sleep* at home on their own terms via digital audio, the piece nevertheless inscribes into its listening reception a certain normative experience of sleep via the temporal logic and implied social etiquette of its performance aspiration. In that sense, *Sleep* merges the aesthetics of the concert hall with the technics of the sleep laboratory by imposing a universal, homogeneous soundscape in which sleep must be objectified and technologized in order to be experienced.

In another sense, the actual performance conditions of *Sleep* reproduce a slightly different environment, namely that of global business travel. A reviewer of the 2016 performance of *Sleep* at the Sydney Opera House noted the "first-class flight touches" like "slippers embroidered with Richter's name, snacks, and eye-masks" that were laid out on her cot in advance of the performance.[17] Despite being a musical respite from the frenetic world, *Sleep* stages the ambience of an overnight flight in business-class, reproducing a concert-hall version of what Sarah Sharma has called temporal architecture: built environments (like jets, airport lounges, and luxury hotels) which fetishize sleep and rest as tools of time-management and work optimization (2014: 43–51). *Sleep*'s atmosphere of work-centered luxury extends even beyond the concert hall. In each of the two outdoor performances in Los Angeles' Grand Park, 500 paying listeners were allowed to break city ordinances and sleep overnight in the park ($60 for a spot on the ground, $80 for a cot), while an at-that-time estimated 58,000 unhoused

Angelenos bedded down somewhere else for the night, probably out of earshot of the performances.[18]

It is worth mentioning that *Sleep* is not the only recent long-durational musical piece meant to be performed or listened to by a sleeping audience. Gascia Ouzunian has created several overnight works that merge live musical performance with live electronics (derived from the EEG data of sleeping performers) and pre-recorded video and audio. Pieces such as "Eden Eden Eden" (2009) and "Music for Sleeping and Waking Minds" (2011–12) offer interactive listening experiences in which audience members are explicitly incorporated into the performance via formal pre- and post-performance discussions with the composer and the performers. The pieces are conceived of as open-ended, experimental forms of communication in and through the media of sleep and sound, and so the intersubjective element is crucial to the concept's execution.[19] Keith and Mendi Obadike have created an 8-hour digital piece titled "lull: a sleep temple" (2021) that fuses aspects of guided meditation with instrumental, electronic, and environmental sounds processed into durational soundscapes. The piece was created and released at the height of the Covid pandemic, so its concept forgoes from the outset the possibility of live performance for a sleeping audience, focusing instead on curating restful, sleep-inducing sounds for the isolated, interior spaces many were forced to inhabit throughout that time. As a sonic invitation to rest, "lull" is also an explicit response to the political and social "unrest" caused by the global protests in the name of Black Lives Matter in 2020, and as such, connects to African American traditions around rest and dreaming as preparatory stages of collective action.[20]

The emphasis on rest as an audio-imaginative preparation for action, not on rest as a superficial pause of 24/7 productivity, distinguishes "lull" from *Sleep*, which follows in the tradition of post-Cagean musical aesthetics that I have discussed throughout the book. As such *Sleep* is marked by the paradox that its transformation of what counts as a musical performance displaces a rarefied, immersive mode of listening onto different kinds of objects or situations, without necessarily critically reflecting on the assumptions and omissions that constitute that mode of listening in the first place. *Sleep* attempts to tune into an ecology of shared planetary rhythms that offers the possibility of rest and a different pace of life, but ultimately tunes into a neoliberal soundscape that presents sleep and listening as purchasable, individualized experiences to be consumed on a 24/7 scale.

6

Listening from Outer Space

Sun Ra's Reverberant Geology

Soundscapes mediate between imaginative, sensory, social, and geologic worlds as an ongoing set of interrelationships that I call geosonics. Throughout the book, I have tuned into aspects of geosonic mediation through an analytical practice of transductive listening, attending to the technical and epistemological infrastructures of the experience of a soundscape. Soundscape is a heterogeneous term in *Geosonics*. More than just the acoustic environment of a particular site or place, I use soundscapes to engage technical processes that sample or sonify site-specific sound or data; immersive, durational compositional practices that elicit a range of attentive and affective engagements across social and geologic space; and the equally immersive, (in)audible ideological environments of global capitalism and neoliberalism. Listening transductively *through* soundscapes means exposing the imaginative assumptions and technical underpinnings of our immersion in sociality and geology, relating lived immersion to technical immersion in/on the earth as the ultimate environment. In turning now to the "interplanetary" music and writings of Sun Ra, I argue that we engage a soundscape in which the imaginative power of sound to transduce the environment ends up transforming the entire planet.

Ra's project is Pythagorean in scope, which means it stretches across the cosmos, but also across the gap between one musical note and the next. Ra was once asked about the role of quarter tones in his music (i.e., tones between the intervals of the Western, 12-tone musical scale), and he affirmed that he could play quarter tones on the piano. He said he had a special technique of attacking the keys, blending harmonic overtones and undertones by touch, to produce quarter tones (Szwed [1997] 2020: 240). Strictly speaking, unless a piano has been de-tuned or prepared in some way, this is impossible. But playing music (like listening to it) is a technical *and* imaginative act. So if the player imagines he is activating quarter tones through touch alone, then who's to say that his imagination does not contour the acoustics in some unheard-of way and allow listeners to (imagine they) hear the piano's unplayable sounds? Maybe we just have to take Ra's word for it. A certain speculative-critical insistence on Ra's word(s) informs my approach to his music, writings, and the challenge to listening that his work provides.

One word that Sun Ra placed special emphasis on was birth. Famous for obfuscating the details of his own birth (and for casually claiming that he came to earth from outer space or from a distant time), Sun Ra expressed the problem of birth not just conceptually, but through the word itself. John Szwed recounts that Ra "would say that birth was the beginning of death, a 'berth' being a place for sleeping, to be 'be-earthed' was to be buried" and that, in one interview, Ra said "'[a] true birthday is the day of your death'" (6).[1] What seems to start out as a play on words (birth, berth, be-earthed) evolves for Ra into a formal philosophical and pedagogical statement: "'And this is exactly what I want to teach everybody: that it is important to liberate oneself from the obligation to be born, because this experience doesn't help us at all. It is important for the planet that its inhabitants do not believe in being born, because whoever is born has to die'" (6). Ra's reasoning, or "equating," is extraordinary. One doesn't counter the inevitability of death with a poignant philosophy of life: one denies birth itself, categorically, and this frees up the capacity to rethink life without death, or birth.[2] By denying birth, Ra denies the limitations of the human body and mind, as well as the temporal arc of human mortality, positing himself as an enduring, non-earthly—and therefore celestial—entity rather than a finite life-form. Ra continues, "'I've separated myself from everything that in general you call life. I've concentrated entirely on the music, and I'm preoccupied with the planet'" (6). On the other side of life (and birth and death), Ra finds only music and the planet, a musical metaphysics that consists ultimately of sonic energies apprehended at cosmic magnitudes. In this highly compressed form is the core of what I call Ra's astrosonics: his scaling-up of human, terrestrial music-making and listening to an inter-planetary level. In this chapter, I show how Ra's work invites us—sometimes sincerely, sometimes playfully, sometimes out of frustration and despair—to listen on a planetary scale, and how the alienness of an outer-space musical aesthetic turns out to have substantial conceptual grounding here on earth, in the technical and imaginative transductions I have followed in the preceding chapters.

In the first part of the chapter, I examine the cosmic scope of Ra's poetics of *equation*, moving selectively through his extensive poetic writings, to sound out the geosonic infrastructure of his musical aesthetics. I argue that "earth" (both as a word and a place in space) is a crucial element in Ra's outer-space imaginary, and that any consideration of Ra's cosmic flights is unthinkable without first addressing his unique reckoning with the earth. In the second part, I turn to Ra's political broadsheets of the 1950s to show how Ra fits race into the planetary equation, transducing a pathway for Black utopian thought on a white supremacist planet. Ra emerges as a Black vernacular Pythagorean, retuning the cosmos by rewriting, one word at a time, a racist history that does not account for Black people, especially African Americans.[3] Parts I–II are primarily *literary* in their analysis, tracing feedback loops between the sound of words and their written forms. Ra's verbal equations demonstrate how meaning transforms itself through sonic and textual repetition, spiraling onwards, even when it seems to be returning back to itself. In the third part of the chapter, I apply the literary motif of textual-sonic spiraling to a particular technique and moment in Ra's music making with his band, the Arkestra, focusing on the heavy use of reverb and delay in their recordings of the early 1960s. What emerged as an accident in the recording process and has been conventionally interpreted as Ra's brief, proto-psychedelic excursion becomes, I argue, a crucial way to listen to Ra's outer space music in the inner space of the recording. Ra's reverb enacts the always-again structure of geosonic listening that attends to the materiality of sound and its reverberation through the planetary imagination.

Part I. Words on Earth: Sun Ra's Poetic Equations

Sun Ra's creativity is articulated through *equations*: connective, associative threads between words and sounds. In conventional mathematics, equations are symbolic statements that describe relations of equivalence. An equation with an unknown element can be "solved," that is, rewritten to define the unknown in terms of what is already known. Mathematical equations are thus static and reductive, not productive. For Ra, however, equations are verbal propositions that speak explicitly to the transformative oscillations between sounds and words, describing conceptual and poetic feedback loops between what a word means, how the word sounds, and the phonetic and semantic slippages of Ra's idiosyncratic etymological play as words/sounds repeat and reverberate on the page and into space. Words are elastic for Ra, stretching, de- and re-forming meaning in ways that might be grounded in everyday speech; in Biblical exegesis; in mishearings, misspellings, or slips of the tongue; or in the intricacies of Ra's racial counter-history and Afrofuturist mythology.[4] In the poem "Words and the Impossible," Ra asserts: "The elasticity of words / The phonetic - dimension of words / The multi-self of words / Is energy for thought - If it is a reality. / The idea that words / Can form themselves into the impossible / Is through the words" (Sun Ra 2005: 431). Words are Archimedean points that serve as the medium or leverage point of their own propagation, their polysemy or "multi-self" stretching to accommodate the "energy for thought." This energetic propagation becomes an equation as it moves across speech and writing. In the poem "Fast Fading Echoes," Ra describes a futuristic vision that "spoke of cosmic equations / the equations of sight similarity / the equations of sound similarity / a secret code of eternal elasticity / clear only to those meant to live beyond / the law of earth" (Sun Ra 2011: 104). Equations mediate the "sight" of speech (writing, textuality) and the sound of speech according to a "secret code of elasticity" that holds tension across the similarities (*not* identities) between words. Equations between words pull us back to a supposed originary meaning, while producing an unexpected set of results at the same time. Words for Ra function like transducers, converting verbal energy across textual and sonic states, creating interfaces that seem both arbitrary and pre-determined, imbuing the "obviousness" of a word, sound, or space with a hidden history and tectonics of shifting meanings, opening pathways for creativity.

Earth, I argue from the perspective of geosonics, is one such key word in Ra's poetics, functioning as the verbal lever that moves sound and listening into outer space. In Chapter 1, I already discuss how Ra's poem "Earth Is A Hole In Space" presents earth as a void that interrupts what would otherwise be the cosmic continuity of space itself. The "earth principle" violates the "space principle" as a black hole (i.e., earthly matter in a state of implosion), a paradoxical emptiness in the fullness of outer space. Yet, sounding out that thought one step further, the void of the earth becomes substantial after all: "the hole of the earth / is the whole / because the earth is surrounded / by the walls of space." Emptiness (hole) reverses into totality (whole), and the finality of the void becomes the fullness of a starting point: "planet earth needs a new idea / another way of approach."

The 1972 poem, "The Empty Space," expands this thought and offers one take on what this new mode of approach could be:

The airy heaven is the empty space
The division of heaven is inner and outer
The limited heaven is the inner
The outer heaven is endless heavens realm . . . . .

The third heaven of the heavens is called earth
It is heaven # three
The earth is as the beginning
It is the foundation of itself
It is an o
Or a o
Which is the alpha omega of the material plane
Yet every planet is an o
But earth is ao #3
Or the third o from the sun.

The inner air from another plane
Is like the counterpoint to a central theme
The enclosed vibration limitation
Repeats itself over and over . . it is a cycle . . an eternity never changing
The outerspirals move brilliantly with word-precision, yet varying
Ever outward and ever onward on.

Sun Ra 2005: 145

The poem situates us in the "airy heaven" of "empty space," but its emptiness turns out to be divided into a "limited" inner space and "endless space" of "outer heaven." The second stanza makes it clear that heaven itself contains smaller units or "heavens" (which elsewhere equate to "havens" for Ra), that is, planets, with earth being "heaven # 3," or the third planet. The use of the number sign in this poetic context is a kind of joke, since Ra is actually counting letters, not numbers. Alphabetically, earth is the third *letter* (in the Greek: alpha [1], beta [2], gamma [3]; gamma [3] = the letter "g" in English, pronounced "gee," and, circling back to Greek again, recalls the Greek word for earth, Γῆ, from which the prefix "geo" is derived). If we count letters not alphabetically, but anagrammatically, and rewrite earth as "erth" (as Ra writes it in the poem "Be-earthed," dating from the same year, freely omitting silent or unvoiced sounds/letters when necessary), then the word becomes an anagram of three, or "thre." So there is a kind of cosmic-literary overdetermination that sounds out earth's place in space via the anglicization of Greek words and letters: earth=erth=thre=3=γ =g (as in geo-)=Γῆ. In a highly compressed form, the phrase "heaven # 3" demonstrates the unpredictable substitutions and transformations of which Ra's equations are capable.

The planet appears in "The Empty Space," as in "The Earth Is A Hole In Space," as a hole, an empty circle, literally a letter "o," but in referring to "an o," or to "a o,"

Ra's poem alludes to alpha and omega, the first and last letters of the Greek alphabet, usually interpreted, via the well-known line from the Book of Revelations, as the enduring presence of God at the beginning and end of all things. The line for Ra implies not only the equation of endings and beginnings, but is also an instance of eclectic or reverse orthography that we see in the equation of "erth" and "thre." Equation-wise, from Ra's esoteric perspective, spelling a word backwards might reveal its real meaning, just as much as a word's sound might resonate beyond its written form, freely jumping across orthographies, alphabets, and languages. There is something rebus-like in Ra's poem, in which an alphanumeric, place-value, picto-phonic textuality (i.e., equation, the "multi-self of words") transduces a relation to the cosmos "beyond the law of earth," yet arising from the word itself, earth. Ra practices verbal geology here, studying the earth insofar as he studies the earth-word (geo-logy), grounding an outer space poetics on the fault lines and sedimented layers of the verbalized earth.

Ra's geology is ultimately conceived of in sonic terms, as we see and hear in the third stanza. The relation between inner and outer heaven qua space, as well as between the heavens themselves qua planets, is structured musically, as "counterpoint to a central theme," and, as it were, acoustically, in which an "enclosed vibration limitation" resonates outward into space as well. It is worth remarking that most musical instruments, as resonators, also act as transducers, transforming "mechanical energy into aerial, vibrating energy through the intermediary of a resonating body" (Augoyard and Torgue 110), which is why a piano tone cannot be reduced to the twanging of a string, nor a saxophone tone to the buzzing of a reed, etc. The hole/whole of the musical instrument transduces the fullness of its sound out of the moving interaction of its parts. A transduction takes place that transforms the partial or limited, inner sounds (or "enclosed vibration limitation") into more complex outward forms. The inner and outer sounds share the same origin and are moved by the same energy (they "repeat [...] over and over . . it is a cycle . . an eternity never changing"), yet their relationship is asymmetrical: the "outerspirals" or outer sounds are altered, "varying / Ever outward and onward on," as the sound propagates into space.

The poem—just like the key words it contains, or even like the keys on Ra's piano that produce quarter tones out of hidden harmonics—is a transducer, or rather a self-transducer, converting its discrepant internal energies into unstable external meanings by way of an oscillating, spiralling pattern that we are invited to imagine as, simultaneously, sonic and cosmic. The equations that articulate these transductive conversions are not logical (in fact, they are dizzyingly circuitous and arbitrary), but they are *precise*. It is "word-precision" that characterizes the transductive movement from inner to outer, that turns repetitions into outward spiraling, and that locates earth in space, us on earth, and transforms earth into a sound that resounds, spiraling into space. The repetition in the last line of the poem ("Ever outward and onwards on"), emphasizes the power of the "word" (qua "out*ward* and on*wards*") as much as it stresses spatial-temporal "on-ness." Being on earth implies a sonic movement onwards/on words into space.[5]

Part II. Factoring Race into the Equation: Ra as Black Pythagorean

There is an inevitable mysticism to Ra's equations, in which a stretchy etymology and an esoteric sensibility are prepared to find secret sonic connections between an inner circle of words and the outer reaches of the cosmos. In terms of aesthetic sensibility, Ra seems to belong as much to an illustrious group of "modern composer-mystics which included Ives, Schoenberg, and Stockhausen, to name only a few moderns; and Ornette Coleman, Cecil Taylor, and Anthony Braxton within the jazz tradition" (Szwed [1997] 2020: 387), as to the ancient Pythagorean tradition that linked heavenly and human bodies via a mathematical epistemology of musical tone. In the famous origin story, Pythagoras, while passing by a busy forge, was captivated by the sounds of five differently sized hammers being struck simultaneously, leading him to discover mathematical ratios inherent in their harmonious overlap (what became the octave, fifth, and fourth) and would eventually reveal the numeric ordering of the universe.

Several contemporary scholars have critically revisited that origin story to account for all that went *unheard* when Pythagoras first hearkened the music of the spheres, emphasizing the selective nature of the mystical hearing that understands itself as tuning in to hidden and celestial sounds. Daniel Heller-Roazen has traced the philosophical incommensurability of mathematics and musical tone at the heart of Pythagorean origin story, which requires that Pythagoras tune *out* of the sound of the fifth hammer, in order to calculate a rational cosmology based on the numeric and harmonic consonance of the other four hammers (2011). Frances Dyson, following up Heller-Roazen's work, expands on the denial of materiality (qua work, noise, and environment) in this origin scene, arguing that the Pythagorean epistemology of musical "tone and its intervallic relations *colonized* the earth and cosmos alike" (Dyson 2014: 22, my emphasis added) as a kind of philosophical and religious pre-cursor to neoliberal extractivism. Likewise, Daniel K. L. Chua and Alexander Rehding read the Pythagorean forge scene as a downplaying of the complexity of music's materiality (2021: 32–7), which in their view unfolds without the need for human agency or understanding. What unites these three revisions of the Pythagoras story is a geosonic infrastructure that subtends the music of the spheres: to listen past the noisy materiality of the embodied and emplaced act of listening in order to imagine the immediacy of universal tones means to ignore the cultural and medial transductions that actually and actively situate the listener in time and space.

Building on the analysis of the previous section, I want to suggest that Ra's equations work both with and against this Pythagorean, mystical tradition. The relation between the cosmos and music established by mathematics (that is, Ra's verbal equations) would situate us firmly in a Western and Pythagorean tradition, if not for the fact that Ra's equations are explicitly founded on the fundamental *inequality* of Black people within that tradition. Indeed, it is arguably the unassimilable histories of slavery and segregation—in Ra's own lived experience, and as reflected in his extensive readings of Black history—that motivate the impulse in his writings to make blackness *equate*, even if it is not possible to make Black people *equal*. In this section, I turn to Ra's political and religious writings of the 1950s where Ra first begins to take race and racism seriously as material for a mythic, Black utopian vision.[6] In this these writings, as in his poetry, Ra's logic is extreme: rather than try to transform an exclusion into an inclusion, to center what was marginal in a positivist discourse that re-affirms a silenced history, Ra intensifies the annihilating gestures of Judeo-Christian and Eurocentric attitudes toward Black peoples with a mind-bending analytical intensity. It would seem that, according to Ra's logic, the violence of such discourses cannot be addressed or corrected through dialogue as much as it must be made to implode on its own brutality. Though the idiom of the broadsheets differs significantly from the more abstract poetry, with the former taking up the rhythms and phraseology of everyday Black speech (including allusions to religious, folk, and popular music), Ra appears to be experimenting with the same the verbal manipulations (repetition, rhyme, wordplay, etc.) that I have already discussed in his later poetry. His particular mode of addressing race and Black history positions Ra, I argue, as a kind of Black vernacular Pythagorean, attempting to accommodate the dissonance of racism into a musical-cosmic equation. Indeed, those aspects of Ra's thought and aesthetics that will later be celebrated as fundamental to Afrofuturism emerge initially, I suggest, as his Pythagorean attempt to retune a utopian soundscape by reimagining collective and everyday Black identities.

William Sites characterizes the broadsheets as a combination of "biblical hermeneutics, millenarian warnings, and religious commentary of various kinds with Africanist mythology, creative etymology, numerology, and pronouncements on the Cold War" (2020: 93).[7] There are obvious parallels and cross-influences among Rastafarianism, the Nation of Islam, and other kinds of Black nationalist and pan-Africanist thought, all of which offered African Americans of the mid-century visionary alternatives to the realities of segregation and the liberal, secular aims of the civil rights movement. By delineating what Ra understood to be a long-suppressed, syncretic origin of Black peoples across Ethiopia, Egypt, Israel, and India, the broadsheets reinvent African American identity through their complex historical rewriting of time and space, and ultimately language itself.[8] But this reinvention must first pass through a bizarre and contradictory rhetoric of self-annihilation, which I trace here in the two-page broadsheet titled "johnny one note." In my analysis of the unique idiom of the broadsheet, I reproduce the idiosyncratic capitalization and punctuation of the original. The vast majority of these texts are printed in all caps (which appears anachronistically as TEXTUAL SHOUTING to the contemporary reader), with ellipses (ranging from two..to more than a dozen. . . . . .dots) sprinkled liberally throughout, and only occasional line or section breaks. The effect is of a single, urgent sentence whose seemingly disparate parts are hammered together phrase by phrase, simultaneously incoherent and eerily cogent.

Riffing on the title of the popular Rogers and Hart song "Johnny One Note" (about an opera singer who humorously derails performances by only singing one note), Ra plays with the idea that "JOHNNY ONE NOTE'S LAST NAME IS DOE. . . .HE'S JOHN DOE. . . . (DO. .re. .mi. .fa. . . so. . .la. . .ti. .do) THAT'S A HELL OF A NOTE ISN'T IT" (105). Johnny's one note is "Do," the first note of the scale in the Western solfeggio system (which teaches singing and ear training by assigning syllables to musical tones). Indeed, there is something solfeggio-like throughout Ra's creative re-ascription of sounds and syllables on shifting linguistic and conceptual scales. As John Do(e), Johnny One Note is an anonymous dead man, hence, for Ra, a Negro/necro (based on Ra's spurious etymological linking of "Negro" with the Greek "necro" in other broadsheets),[9] and a monotonous and meek figure ("JOHN PLAYS DOE ALL THE TIME. ISN'T HE A DEER..HE PLAYS DOE ALL THE TIME...MONOTON OUS ISN'T HE" [105]). Elsewhere in this broadsheet, Johnny One Note is linked to another vernacular phrase, Johnny-on-the-spot, which conventionally means a reliable and ever-present helper or friend, but which Ra recasts as a sign of spiritual stagnation: "YOU'RE IN A SPOT...JOHNNY ON THE SPOT TO BE EXACT..YOU'RE IN THE SAME SPOT YOUR ANCESTORS WERE 3,000 years ago. YOU HAVE MADE NO SPIRITUAL PROGRESS SINCE THEN. YOU HAVE BECOME THE LAUGHING STOCK OF THE WHOLE WORLD" (104). The monotonous collective ("monoton ous," or *monotone us*) of the contemporary Negro/necro is associated with an ancient condition of cultural and spiritual shame. Ra's text takes the shape of an angry sermon, provoking its intended audience (the African American community of 1950s Chicago) by mobilizing anti-Black stereotypes in a partially biblical, partially modern milieu.

As Ra's tirade spirals onwards, he uses the "one note" motif to equate ancient Black and contemporary African American identity with zero, in social and quasi-mathematical terms. As with "heaven # 3", the reasoning operates at the edges of intelligibility, echoing with a broken-telephone-like effect through a range of staggering permutations: "JOHNNY IS NOTED...HE'S RENOWN..HE'S FAMOUS...HE'S FAMED..HE IS OF NAME...HE IS THE FAMED NAME...HE IS THE NAME...THE NAME IS SHEM...JOHNNY ONE NOTE IS SHEM [...]" (105). This passage hinges on the tautological reversal of the phrase "he is the name" into "the name is him," only Ra contracts "is him" to a single sound/name "Shem," evoking the son of Noah from the Old Testament. Shem functions as another phonetic and associative variable in the equation for Ra, moving toward the astounding conclusion of the broadsheet's first section:

> SHEM IS JOHNNY ONE NOTE.....HE'S I0..I0 is one 0....one o is zero....zero is nothing...SHEM is nothing. NEGROES ARE NOTHING....there is no such thing as a negro therefore negroes are symbolical of nothing [...] NEGROES ARE SHEM. NOTHING IS MYTH.....HAM IS A MYTH...NEGROES ARE HAM.......*theres a common expression "A nigger ain't—. Less than that is nothing. Zero* (105)

Shem, in the flow of Ra's thought, equates both to Ham (Shem's brother in the Old Testament) and to the word "sham" (e.g., a hoax, trick, fraud, etc.). The broadsheet reinterprets African Americans (through the equation of Johnny One Note and Shem) as a cipher of nothing through a typographical play on the first two letters of John rewritten (presumably according to the Latin i/j equivalence) as Iohn, then respelled, partially and phonetically as I-Oh, and then Roman-Arabic-numeralized as I0 (one zero). Shem (qua sham) signifies the non-reality of a fake, an association weirdly confirmed by equating Johnny One Note with zero, nothing. By repeating, varying, and returning to the same phrase—and the same associated cluster of words, names, and sounds— "johnny one note" effects a rhetorical paradox of multiplicity within nullity. African Americans (as summed up in Ra's quoting of the verbally annihilating n-word) are *not*; equation-wise, they equal nothing. But, at the same time, that nothingness is distributed across a proliferation of places, names, and identities that fill the putatively empty space of Black culture, from the ancient world to the present, with a sense of fluidity and incipient transformation. Sites reads this rhetoric as a kind of pedagogical estrangement: "At any moment, these [broadsheet] texts imply, African Americans may become alluringly different from how they so often see themselves, taking on new identities full of possibility, transcendence, and moral agency" (2020: 126). Paul Youngquist observes that the broadsheets "dislocate[] American Black identity from whiteness as its defining condition" and his commentary on "johnny one note" notes that "Sun Ra will exploit this empty status [the zero] in his later poetry, myth, and music, attributing positive content to 'nothing' that associates it with infinity" (2016: 57, 59). The harsh equation "NEGROES ARE NOTHING" thus loses its valence by the sheer proliferation of other variables that rewrite the equation and may eventually revaluate it altogether.[10]

"johnny one note" accomplishes a similar paradox as "The Empty Space" insofar as both texts defamiliarize their subject (African Americans and planet earth, respectively) by razing our conventional, historical conceptions and filling the void with a potentiality already in progress. And, just like the poem, the broadsheet conceives of that potentiality in sonic terms. The second page of "johnny one note" transforms the obstinate, singular "note" from a word to a sound:

> ONE NOTE IS A SOUND ...negroes are sounds... witness the expression, "HOW YOU GON' SOUND, MAN?".... "HOW YOU GOIN' TO SOUND, MAN?" / if negroes are not Johnny one note, WHY DO THEY ASK ONE ANOTHER, "HOW YOU GOIN' TO SOUND?...... / i know how i'm going to sound. i'm going to sound so loud that it will wake up the dead.
>
> (105–6)

Sun Ra takes an everyday colloquial expression ("How you goin' to sound?") as proof of (sonic) life. Only after Johnny One Note is pronounced dead and the African American is reduced to zero can the other side of the equation be properly heard: "negroes are sounds." The broadsheet posits a vitalism of blackness as sound that condenses and transforms the multifaceted nihilistic cipher of the Negro/necro. We have to note the significant shift to an active and first-person voice (rendered oddly emphatic by the rhetorically understated diction and lower-case typing): "i know how i'm going to sound. i'm going to sound so loud that it will wake up the dead." Sites notes that the phrase "it sounds so loud as to wake to the dead" comes "from 'Where Shall I Be,' a popular hymn written by turn-of-the-century Black Holiness preacher Charles Price Jones" (122). The phrase thus adds Christian music to the pop-musical clamour that reinforces, but ultimately reveals and transforms Black alienation in the broadsheets. The passage adds an additional twist on its sonic vitalism: "SOUND also means FOUNDED ON TRUTH OR RIGHT. / ARE NEGROES SOUND? EVERY NEGRO WHO LOVES GOD IS SOUND. / sound means founded on truth or right. EVERY NEGRO WHO LOVES GOD IS SOUND" (106). The repetition of "EVERY NEGRO WHO LOVES GOD IS SOUND" adds a redemptive moral dimension to the double-entendre: if Negroes are sound(s), then Negros are sound(s). The broadsheet concludes with more conventional Christian exhortations to choose life and glorify God, to turn away from the evil and moral shame (etc.), but it is clear that the transformative moment, stylistically and conceptually, comes with the introduction of—and agency associated with—sound, which both interrupts and flows directly out of the discordant "Johnny One Note" refrain. By riffing so obstinately on a single phrase, Ra positions himself and his audience/readers not just as conductors and listeners of sound, but makers of sound who analyze, interpret, and create through repetition.

In Part I, I suggested that earth is a key word (an "on word") for Ra's poetic cosmology, arguing that that Ra enacts a geosonic imaginary to conduct his spiraling excursions into outer space. I want to conclude Part II by sketching briefly how the hard-fought equation of blackness and sound from the broadsheets connects forward in Ra's thinking both to earth and to outer space. Ra's 1972 poem "The Sub-Dwellers" presents a sonically transfigured blackness at the core of a cosmic vision. The poem begins: "Down in the subterranean places of the city / Down in the catacombs and caverns of the mind / Down, / Down in the earth catastrophe of knowledge / Dwell they, the sub-dwellers . . ." (2005: 362), designating a marginalized population forced literally and metaphorically underground. As the poem continues, the repetition of "down" transforms into a "dawn" that will "touch them with immortal rays," and here the poem mixes imagery of burgeoning roots and seeds with the bursting forth of sound: "They [the sub-dwellers] shall come forward to the sound of sounds / Yes, a sound of sound will burst the oath of earth asunder." The poem concludes with a preacherly flourish:

> And that unmeant secret place will be no more their home
> For out of the earth-darkness
> They shall come forward dancing to the sound beams of dark rays of light
> And they will rise to the heavens of spiritual skies
> There they will be like the fires of the rays
> And the rays of the sound of the sun.

The complex imagery invites us to read the mixed metaphor of light/sound as it simultaneously radiates down to earth from the sun and bursts forth out of the "earth-darkness" as solar rays. The subdwellers are transfigured as "the rays of the sound of the sun," as a vibratory energy that takes the form of both light and sound. "Rays" is nearly a homophone of "race," an association that Ra implies in "The Sub-Dwellers," but makes more explicit in other texts. His 1968 essay, "My Music is Words," contains the statement, "My measurement of race is rate of vibration-beams . . . rays . . . Hence the black rays is a simple definition of itself/ phonetic revelation" (Sun Ra 2005: 467).[11] And the undated piece "The Black Rays Race" is a taut tautological poem which attributes to the "Black race" an immeasurable, but misunderstood wisdom that transforms and manifests itself as the "unlimited freedom of the black rays" (83). What I am pointing toward here is Ra's translation of his earlier exegetical/cabbalistic equation "negroes are sounds" into the cosmic Afrofuturist duality of blackness as a race and as solar rays, reimagining a marginalized, segregated, metaphorically underground people and history as a utopian and cosmic force. The sound that wakes up the dead (i.e., Negro/necro) in Ra's earlier Biblical paradigm is rendered as cosmic rays/race in Ra's outer space paradigm. This puts a spin on the late 1960s notion of a "space race": punning on a phrase that depicted the two Cold War powers trying to outdo each other in the colonization of space, Ra's space race gestures toward a shift in the cultural perception of Black peoples (especially African Americans) through a shift in how we attend to words ("phonetic revelation"). On this point, Brent Hayes Edwards has noted that, "[g]oing to space is epistemological work—it might force us to alter our conception of what 'the inhabitants of this planet' can be. It 'races,' but more *razes* and *raises*, as Ra might say, the potential of the human" (2017: 125), while at the same time pointing specifically to the exceptional status of African Americans as terrestrially marginalized, but cosmically central.

Returning to the question of Ra as mystic (as a Pythagorean who spells out the meaning of the cosmos by listening to the sounds of words), we can see here the evolution of that earlier attempt to make racial inequality equate—to think the nonbeing of the African American as it is evoked in vernacular speech and song lyrics, as well as born out in the lived experience of Black people under segregation. Ra is a Pythagorean with a crucial difference: the harmony of the world is not perceived by abstracting a cosmic ratio out of everyday experience. It is created by taking the elements that have been forced out of the equation and buried in the earth, and rewording and resounding them until they make a new kind of sense as an imaginary world or a utopian possibility. Ra's insistence on words as Archimedian points of cosmic leverage is grounded in his geology of the earth as the medium that can project a misunderstood, marginalized Black utopian vision into outer space. Ra as Black vernacular Pythagorean sounds an imaginary cosmos through a down-to-earth Black soundscape.

Part III. Re: Words, Reverb

What remains difficult to imagine, in general, is how Ra's poetics would translate into a musical practice. What would a verbal equation sound like as music, and how might these equations help listeners to understand specific aspects of Ra's approaches to composition, improvisation (including "conducted improvisation"), and recording? Finally, how, if at all, can we convert the preceding argument for Ra's geosonic poetics into some musical practice or corresponding technique of listening that we could imagine as a sound(ing) of the earth?

Ra's writings fold sound in and out of language, and position the act of reading intermedially as a kind of listening (often a mis-hearing) that makes meaning in the interstices of a word's visual, sonic, and social texture. Yet these same writings might not necessarily help us as listeners when confronted with the actual sound of Ra's music. In some of Ra and the Arkestra's most daunting, experimental works (e.g., albums of the mid-60s "New Thing" moment, like *The Magic City* and *Heliocentric Worlds*), there can be ferocity and desperate searching intensity to the music which belie the oracular placidity of his poetry. And the writings' mystical and cosmic sweep would seem to contradict the mind-boggling diversity of Ra's musical output, which alongside the experimental jazz, "space music," and "cosmo drama" of the Arkestra's live performances, includes a lifelong adherence to the traditional big band sound (via his collaboration and many subsequent homages to Fletcher Henderson); Disney tunes; Gershwin and other Tin Pan Alley standards; and even some session work on a Batman-and-Robin-themed commercial recording. Ra's vast catalogue as composer, bandleader and performer (which includes hundreds of recordings, and is still growing as unknown or rare recordings are uncovered and re-released) could keep an entire team of music scholars busy for a lifetime.[12]

To sound out one possible link between the equations and the music, I want to focus on the very specific use of audio delay that Ra used in a number of recordings of the early 1960s. I will concentrate on a single track, "The Voice of Space" from *Cosmic Tones for Mental Therapy* (1963), though the effect appears on several other recordings from that period.[13] The delay effect began as an accident in the recording process. In the early 1960s, the Arkestra was rehearsing at the Choreographer's Workshop at 414 West 51st Street. The band's second drummer, Tommy Hunter, had acquired a pawned Ampex tape recorder, at Ra's request, and was tasked with recording and editing the band's rehearsals. Szwed recounts that, while testing the recorder at the start of a rehearsal session, Hunter accidentally "discovered that if he recorded with the earphones on, he could run a cable from the output jack back into the input on the recorder and produce massive reverberation" ([1997] 2020: 187). Hunter himself recalled: "'I wasn't sure what Sun Ra would think of it. . . .I thought he might be mad—but he loved it. It blew his mind! By working the volume of the output on the playback I could control the effect, make it slow or fast, drop it out, or whatever'" (187). Szwed calls the effect "reverberation" (conventionally shortened to "reverb"), while other critics refer to this technique as "Hunter's echo machine" (Coleman 2002: 4).

The three terms—delay, reverb, and echo—are technically distinct, but all gesture towards the "always again" structure of listening that I discuss in the Introduction and Chapter 1. Delay is, strictly speaking, a feat of signal processing that repeats a sound over itself, at a slight temporal lag, to create the repetitious effect of a sound both cutting itself short and prolonging itself at the same time. Delay interrupts sound by repeating sound, and vice versa. Depending on how the delay is manipulated, it can produce a feedback of subtle ripples within a given sound, or generate massive sheets of noise. "Reverb" is a specific kind of delay that avoids interruptive feedback effects and produces a sense of spaciousness in an otherwise "dry" or "dead" signal. Reverberation, as I mention in the Introduction, is also a more general acoustic term that refers to all the ways in which a sound fills a given space, bouncing off all the available surfaces within it. Reverberation refers to the acoustic merging of sound with the sounding of the space in which sound occurs, hearing the two together (again) as a composite, spacious sound. Finally, echo is a certain type of acoustic reverberation in which we hear a given sound repeat back more or less distinctly as it reverberates off a given surface. Taken together, these distinct audio and acoustic terms refer to a kind of meta-listening, or what I have called the "always again" aspect of the act of listening that converts sound back and forth between a material process and a culturally imagined meaning. With delay/reverb/echo, we hear the "signality" and spatiality of sound, and the temporality of our own listening process.

Ra's use of delay is undeniably "spacey" and is usually acknowledged as an experiment in "proto-psychedelia."[14] I want to work against that interpretation. The psychedelic angle implies that Ra would use reverb to defamiliarize conventional listening habits, manipulating signals to create an altered, hyper-acoustic inner space of listening. I argue, on the contrary, that Ra uses reverb not to invite us to listen differently, but to reinforce what listening already is. Ra's reverb repeats back to us how listening works as a transductive process of equation making: converting unknown variables (new, unheard-of vibrations and signals) into known ones (imagined or understood sounds), and likewise, turning familiar, culturally known sounds (the sound of a saxophone, of a piano, of "jazz") into material and metaphorical unknowns. In Ra's music, the feedback loops caused by the interference between knowns and unknowns merge with the listening process itself, as it simultaneously tunes into itself, and spirals out into radical alterity. An illustrative moment comes in the aforementioned "Voice of Space" from *Cosmic Tones*. The piece begins with Ra on the organ (probably the Hammond B-3 borrowed through Hunter's connections and used on other recordings between 1961 and 1964), playing long, piercing trebles notes against quickly fingered passages in mid-bass range. Doubled (and then some) by the heavy use of Hunter's reverb, the flurry of low organ notes creates a murky, yet propulsive rhythm-like effect that hums in the background throughout most of the track. Ronnie Boykins enters with heavily accented bowing on the double bass, a repetitive pattern (if never quite repeated in the same way) of descending intervals, well up the neck, with a supple, almost cello-like tone. The only percussion we hear is the regular/random, drum hits (Clifford Jarvis) that ping through the reverberant texture of the organ notes, marking a time that does not seem to coincide either with Boykins' bass playing, or Ra's stabbing treble notes. The overall sound of the piece is incredibly busy, full of activity, yet without a clear rhythmic or tonal center (i.e., performed in the fluid tonality that Ra often called "space key").

Curiously, the sonic space is dominated more by the delayed notes than by the "actual" playing of the musicians. This paradox of unplayed, but audible musical time is enhanced with the entrance of Danny Davis on alto saxophone. His surging, ascending riffs cut off abruptly, deliberately anticipating and playing into the ensuing delay that would be audible on the recording. At two and a half minutes, the reverb is intensified to produce a crescendo of hissing feedback (I would call it "space tinnitus"); the organ and saxophone drop out, while Boykins plays on in frantic tremolo. He abruptly shifts to a jagged, ascending riff into the upper harmonic reaches of the instrument until he literally runs out of space on the strings and is cut off by the now surging, whirring sheets of feedback. There is an indeterminate, transitional moment of hissing space tinnitus, then the organ enters and takes over with both hands churning away insistently in a fugue-like passage. By five and a half minutes, the organ has dropped out again, and now the saxophone (like the bass earlier) plays frenzied, exposed solo runs at the upper reaches of the register, overlaid with—and at times completely overcome by—waves of feedback. The urgency of the playing, coupled with the intensity of the feedback, suggests a climax has arrived, and then for a few seconds, something completely ordinary happens, but in the most extraordinary manner. Hunter cuts out the reverb completely (he is actually credited as playing "reverb" on some releases of this recording!), and for the first time in the piece, we hear the "clean" sound of the saxophone as Davis plays a wild, rapid, yet precisely articulated descending riff. The clarity and intimacy of the sound are astonishing in contrast to the cacophonous reverberations that precede it. Suddenly all that spaciness and irregularity contract, and we are just in a room listening to the sound of a saxophone, but as if we've never quite heard it that way before; the sound is almost unrecognizable precisely because it is so "naturalistic." After just a few seconds to catch our breath, Hunter turns the "echo machine" back on, and the piece moves to a coda of pinging drums, tremolo bass, and Hammond organ pulses, all moving thickly through the spacey soundscape that has, by now, become quite familiar, even mundane, to our ears.

At the level of an individual sound, this use of delay makes the end of that sound coincide with its beginning-again. Delay is, in a sense, the audio equivalent of Ra's reverse orthography. Ra's spells words backwards not to defamiliarize their meaning, or reveal some occult depths, but rather to show how a word's true meaning is distributed across a range of possible orthographies and pronunciations. Recording-wise, then, Ra's delay is something quite different from literally playing a recording backwards for distortion's sake, or to reveal an occult dimension (e.g., to learn via murky vocal mumblings that, "Paul is dead," when you play the Beatles "Revolution 9" backwards, as has been popularly attributed to the "White Album" of 1968). Nor is it a case of using reverse playback (and other forms of audio and tape manipulation) toward the sonic sculpting of *musique concrète* in the manner of Pierre Schaeffer, where the inversion of a sound defamiliarizes us so radically to the time and space of its production that the sound "itself" attains aesthetic autonomy. Nor is Ra's reverb exactly an instance of an audio effect taking on a culturally symbolic value, as Michael Veal has argued about the use of echo and reverb in Jamaican dub remixes of the 1970s. For Veal, the Black diasporic relation to (actual and imagined) cultural roots in Africa—and to the complex historical violence of the slave trade that thwarts linear modes of narrative and representation—finds a symbolic expression in the heavy use of "reverberation [to] provide[] the cohering agent for dub's interplay of presence/absence and of completeness/incompleteness" (199). In contrast, I suggest that Ra uses delay to create and superimpose an additional temporal-sonic plane on which the ending of a sound would coincide with its beginning, creating a centripetal sonic cluster that still moves forward in time. These micro-repetitions on the "inside" of individual sounds transform into to the overall effect of sonic spiraling into "outer" space, that is, the reverberant non-space of listening with reverb. In this sense, what we hear in "The Voice of Space" is the spaciousness of our own careful listening as we simultaneously listen "back and forth" to individual sounds *and* to their cumulative resonance in the imagined space of our listening throughout the piece.[15] And yet, in those few seconds where the reverb cuts out, we return to the sound we have been hearing all along, and hear it again, only wrested out of the outer spaces of our reverberant listening and re-situated in the humble, yet mind-blowing space of the sound of the saxophone. In this single track, reverb makes it possible to be on an outer space orbit *and* right down here on earth all at the same time.

The liner notes to *Cosmic Tones* include a programmatic statement from Sun Ra that reads: "PROPER EVALUATION OF WORDS AND LETTERS IN THEIR PHONETIC AND ASSOCIATED SENSE, CAN BRING THE PEOPLE OF

EARTH INTO THE CLEAR LIGHT OF PURE COSMIC WISDOM."[16] But there are no lyrics on this album (unlike many other Ra recordings which feature singer June Tyson; "space chants" sung by members of the Arkestra; even sermons from Ra himself), only small chamber-like ensembles playing improvised, *instrumental* music. Why are we invited to think about words and letters while listening to an album that offers only musical sounds? Perhaps the key is reverb itself as the audio equivalent of Ra's elastic poetics of the word. Ra constantly rewords words to distribute their meanings across their "PHONETIC AND ASSOCIATED SENSE." These are the verbal manipulations that I analyze above in Ra's poetry and broadsheets: the use of repetition to produce difference via homophony and homonymy; inversion, permutation, and numeration of letters; transliteration into other alphabets and vernacular orthographies; and Ra's idiosyncratic techniques (part Kabala, part playing the dozens) of etymological and exegetical excavation. In audio terms, reverb rewords sound; and, if I can make my own slant etymological riff, they are in fact the same word, via the Latin *verbum*: re-word is re-verb. Reverb introduces an element of repetition into the listening process that activates and transforms the time and space around musical sounds. Rather than de-familiarizing and estranging listeners, Ra's reverb re-familiarizes us to new auditory potentials built in to the always-again temporality of listening. This use of reverb allows sounds to circulate more freely in imaginative space precisely because they are also rendered more material as vibrations in a transductive process of becoming. The "Voice of Space," then, is not just a metaphor, nor is it just a title, nor is it a programmatic attempt to translate a cultural myth into sound (though it is certainly all these things, and more). It is a reverberant call that hails the peoples of the earth and summons them toward some as yet unheard-of cosmic wisdom. Spiraling on words/onwards, Ra's reverb achieves escape velocity for geosonics.

Geo to Astro

Stephen Jay Gould's study of the origins of geologic time, *Time's Arrow, Time's Cycle* (1987) includes the following quotation (attributed to George P. Scrope, an eminent geologist of the early nineteenth century) as an epigraph: "The leading idea which is present in all our researches, and which accompanies every fresh observation, *the sound to which the ear of the student of Nature seems continually echoed* from every part of her works, is—Time! Time! Time!" (my emphasis added). Gould notes that "[t]his quote has become a virtual cliché by overuse in modern textbook epigraphs," yet it may gain fresh relevance here by pointing to the congruence between geology, media, and sound that I discuss throughout the book. For theorists like Siegfried Zielinski and Jussi Parikka (discussed in Chapter 1), the earth is a giant inscription medium for deep time, and media devices (their material origins, their energy footprints, their afterlives as e-waste) carve out their own notches, their geosonicks, on that deep geologic timeline. Mediation itself is another process of geosonicks, of making cuts or interfaces (*Schnittstelle*) in the earth (and with the earth-media of rock, water, and air) that stand in some permanent relation of partiality to the imaginary whole. John McPhee, the writer and non-fictional-geologist who coined the phrase "deep time" in the book *Basin and Range*, was also obsessed with road-*cuts*, that is, blast sites necessary for highway construction which, as an added bonus, expose buried layers of rock for spontaneous examination by traveling geologists. Road-cuts, like exposed, weathered rock, tell stories that contribute to what McPhee called "The Picture," the implied, but always partial grand narrative that links the fragmentary data of road-cuts, earth borings, and seismic sounds to a continuous flow of geologic time (1998: 62–3). If Sun Ra is to achieve true escape velocity for geosonics, then he would have to factor not just the earth, but time into the equation.

Playing a version of himself in his 1974 film *Space is the Place*, Sun Ra declares: "Equation-wise, the first thing to do is to consider time as officially ended. We work on the other side of time." In the context of the film, what lies on the "other side of time" is the alien planet where Ra, in his role as intergalactic musical psychopomp, is setting up a "colony for Black people" to "see what they do with a planet all their own, without any white people there." Setting aside the details of the plot—which is partly a retelling of Ingmar Bergman's *The Seventh Seal* in the cinematic language of blaxploitation and low-budget sci-fi, partly a music documentary, and entirely a disarming satire of Ra's whole aesthetic—Ra's interplanetary transcendence of time turns out to be grounded rather humbly in good old-fashioned geology. As I suggest above, geology's ultimate object is not the earth, but time itself. For the geologist, the earth *is* time, or a structuration of time, and it requires a scaling-up (-down? -out?) of human temporal consciousness to an extreme so earthly that it becomes alien. On a geologic timescale, the geologist glimpses humanity from an alien perspective, as "a race unaware of its own instantaneousness in time," following briefly in the footsteps of "all the species that have come and gone" (McPhee 1998: 90). Such insight strains the geologist's ability to think meaningfully about geologic time:

> The human mind may not have evolved enough to be able to comprehend deep time. It may only be able to measure it. [... Geologists] wonder to what extent they truly sense the passage of millions of years. They wonder to what extent it is possible to absorb a set of facts and move with them, in a sensory manner, beyond the recording intellect and into the abyssal eons. [...] On the geologic time scale a human lifetime is reduced to a brevity that is too inhibiting to think about. The mind blocks the information. [...] In geologists' own lives, the least effect of time is that they think in two languages, function on two different scales. (90)

Geologists struggle to integrate the deep time that is their crucial epistemological tool into the "shallow" time of human consciousness and experience. The span of a human life can appear, at best, on the scale of a "race" or "species" whose evolutionary clock has long been winding down, or "liken[ing] humanity's presence on earth to a brief visitation from elsewhere in space" (89). It would seem that Sun Ra had arrived at a similar alien perspective about race, only he worked it out through his distinct combination of Biblical and etymological research; Afrofuturist mythology; and musical experimentation. However, rather than being forced to "function on two different scales," Ra learned to equate them in a single creative language by canceling out the equivalent variables. The equation might run like this: if the end of human time is death, and death is birth (qua berth, be-earthed), then time and the earth are inextricably linked. To cancel death out of the equation means canceling out the "law of earth," which is time. Ra wants to annihilate the need to be be-earthed (to be born, to die, to live according to the law of earth, which is also the law of time), and so he also annihilates the earth itself, all in order to balance the equation. Ra fills the imaginative void where earth used to be with imaginary planets "on the other side of time" (like the alien planet with the space colony in *Space is the Place*). The coincidence of terrestriality and temporality is taken out in a single stroke by Ra's astrosonic inversion of geology, which renders the earth into a "hole in space" and summons an interplanetary soundscape populated by beings liberated of the necessity of birth/be-earth. In an uncannily similar vein, when McPhee cites one of his conversations with environmentalist David Brower on the subject of deep time, it is hard to tell where the geologic stops and the astrosonic begins: "'If you free yourself from the conventional reaction to a quantity like a million years, you free yourself a bit from the boundaries of human time. And then in a way you do not live at all, but in another way you live forever'" (1998: 91). This is geology, but it sounds like Sun Ra.

Afterword

Part I. Afterward

Geology supplies two useful concepts to look back and survey the ground covered in *Geosonics*: unconformity and uniformitarianism. An unconformity is a temporally heterogeneous structure of rock, that is, a single geologic site which neatly juxtaposes strata formed at vast temporal removes on the time scale. The famous unconformity at Siccar Point in Scotland, for example, presents a horizontal layer of red sandstone resting directly on top of a vertical layer of greywracke. Modern geologists know that there is a gap of roughly 60 million years separating the formation of these rocks, yet there they are, plainly touching each other (Knoll 2021). Without needing (or even knowing how) to date the rocks, James Hutton, the eighteenth-century founder of modern geology, recognized in Siccar Point the immense amount of time necessary for geologic processes first to deposit the bottom layers of rock on the seafloor, gradually tilt them upward (while also eroding them), and then deposit yet another layer of different rock altogether on top of that. Siccar Point afforded Hutton's dizzying glimpse into "the abyss of time" (in the later words of Hutton's companion in visiting the Point, James Playfair [Knoll 2021: 42]), but for Hutton the unconformity was also a paradoxical image of stability. Not only was that seam of rock evidence that the Earth was far older than the contemporary Biblical-creationist timeline allowed; it also was, for Hutton, proof that time itself could be an explanatory force of the shape of the world. Without deep time, early geologists had to posit catastrophic events (earthquakes, floods, storms) to account for any otherwise inexplicable rock formation. In Siccar Point, Hutton saw that time provided the necessary scale on which cycles of accumulation, uplift, and erosion could play out, thereby assuring the deep continuity of geologic processes and inaugurating *uniformitarianism* in modern geology (Bjornerud 2018: 24–5). Uniformitarianism stresses that geomorphic forces operate at relatively uniform rates over billions of years, positing holistic, cyclical processes that tie together locally contradictory formations. Yet as Hugh Raffles points out, Hutton's Unconformity at Siccar Point is "both a seam *and* a rupture," marking an interval of time permanently lost from the geological record, even as it documents the deeper temporal permanence of geologic processes (2020: 5, emphasis added). Unconformities, then, attest to locally irretrievable gaps in the geologic record while activating the conceptual tools that make uniformitarian geochronological knowledge possible in the first place.

In *Geosonics*, unconformity is an organizing principle. The deliberate gaps between chapters are missing pieces of a larger narrative that traces a number of overlapping histories:

- the twentieth-century history of geotechnical listening and the weaponization of sound across military-industrial and academic contexts;
- the history of sonification techniques for research and research-creative purposes, especially the aesthetic and musical applications of geophysical data;
- a history of connectivity that conjoins telecommunications, environmentalism, and extractivism in a vicious material and ideological circle; and,
- a history of sonic-ideological interventions in the soundscape that take the centrality of sound and listening for granted, while also constraining the potential of diverse listeners to imagine sound and environment otherwise.

Such continuities are present in the otherwise discontinuous transitions between, say, a chapter on seismicity and a chapter on Beethoven's 9th, or a chapter on sleep and a chapter on Sun Ra—who, by all accounts, did not sleep![1] But an astute reader will, like Hutton, recognize those implied continuities in order to draw into focus the complex, unconforming presentation of the material at hand. The chapters of *Geosonics* are analytical cuts, cross-sections, borings, soundings through an evolving soundscape that transduces the world-making powers of media through the real and imaginative geomorphological agency of humans.

Just as geologists analyze geomorphology into discrete, interacting processes (sedimentation, subduction, erosion, etc.), so does *Geosonics* make several discrete critical interventions within its overall "uniformitarian" approach. *Geosonics* de-essentializes the soundscape concept by attending to the immersive presence of environmental sound as a media artefact with a technocultural infrastructure (Introduction and Chapters 1 and 2). The book also contributes to an expanded notion of media interfaces in critical, technical, and aesthetic/phenomenological terms (Chapters 4 and 5). As an experiment in speculative listening, *Geosonics* imagines that an earth-scale system like 24/7 capitalism, or an ideology like neoliberalism or audism, could make or transduce sounds, and proposes speculative-critical listening practices that attend to these imagined sounds (Chapters 3, 4, and 5). Shifting between the speculative and the material, *Geosonics* discloses the geologic earth as it has been embedded in human auditory imaginations (Chapters 2 and 6) and reckons with the politics of who and what gets heard on the man-made earth (Chapters 4 and 6). And finally *Geosonics* explores analogies (e.g., unconformity) between the instability of words and the instability of geologic processes, revealing how individual words compress distant times and meanings, slipping along phonetic and orthographic fault lines that change the semantic landscape, while preserving traces of unknown origins (Introduction, Chapters 1, 4, and 6, and Afterword).[2] As Christine Sun Kim's and Sun Ra's work show (each in different ways), words are simultaneously signs and sounds, but their simultaneity should not imply a symmetry: words/sounds are inflected by power and environment, and change their meanings as they spiral through communicative space. Tracking the materiality and technics of sound from the human scale of communicative and sensory action to the earth scale of planetary listening is the ambitious conceit of this book. In pursuing it, *Geosonics* follows a nonlinear, but uniformitarian narrative flow that tolerates discontinuities and juxtapositions of disparate material as part of its underlying inquiry into sound's geologic mediation.

Part II. Afterwor(l)d

In *Timefulness: How Thinking Like a Geologist Can Help Save the World* (2018), geologist and writer Marcia Bjornerud expands the uniformitarian view of geology into the basis of an argument for the centrality of geological thinking more broadly. Playing on the wellness term "mindfulness," timefulness comprises an expanded awareness of the complexity of geologic processes, and points to the need for new, sustainable imaginings of the future on deeper time scales than the short-term view of contemporary capitalism and geopolitics. For Bjornerud, "[r]ecognizing that our personal and cultural stories have always been embedded in larger, longer—and still elapsing—Earth stories might save us from environmental hubris" (178) and "[m]aybe, just maybe, the Earth itself, with its immensely deep history can provide a politically neutral narrative from which all nations may agree to take counsel" (19). *Geosonics* contributes, from a sound and media studies angle, to a centering of the geologic analogous to Bjornerud's timefulness, by arguing for the aesthetic and conceptual expanding of the earth as constitutive of how and what we know, especially when we start from the enduring paradox that "Earth has been an elusive subject to study—simultaneously too near and too far away to get into clear view" (23).

I want to conclude, however, by shifting from Bjornerud's uniformitarian concept of timefulness back to its geological counterpart: catastrophism. The unfolding of sudden catastrophic events leaves undeniable imprints on the geologic record, and the task of explaining such events has been as constitutive of geological knowledge as understanding the gradualism of deep-time processes. Geologist Anita Harris, recounting her experience of a massive 1959 earthquake in Montana to John McPhee in *Annals of a Former World*, says, "We were taught all wrong. We were taught that changes in the face of the earth come in a slow, steady march. But that isn't what happens. The slow, steady march of geologic time is punctuated with catastrophes. And what we see in the geologic record are the catastrophes" (1998: 171).[3] Geologic catastrophes interrupt the cyclicality of uniformitarian geology and force scientists to integrate starkly delineated before-and-after temporalities into the gradualism of geologic time.

When human beings attain geomorphological agency (and thereby inaugurate a new geologic epoch, the Anthropocene), they do so *catastrophically*, and thus enact a temporal contradiction: how to measure the here-and-now of the still-unfolding catastrophe in terms of the speculation about longer-term effects on a geologic scale? How to transform a notion of deep time, always grasped implicitly or explicitly as belonging to the past, into a future-oriented way of understanding the present? Geoscientists, climate activists, as well as politicians and industry leaders (most of the latter two with appalling insincerity) are grappling with this new epistemological and existential planetary question. I want to conclude by asking what this question *sounds* like and how we can begin to attend to this problem transductively.

I opened this book with a rhetorical question about the audio equivalent of the *Blue Marble* photograph, and the answer I gave (which helped to introduce the speculative-critical approach to sound, media, and environment of *Geosonics*) in fact obscured the correct answer in space-exploratory terms. The Golden Record aboard NASA's Voyager spacecraft (launched in 1977) contains audio material—greetings in fifty-five languages, samples of music from Western and world folk traditions, and field recordings of natural sounds, animals, and humans—that was meant to be a representative picture of the earth (and its inhabitants, mainly human) in sound. While it is tempting to historicize and deconstruct the sonic essentialism of this bold curatorial gesture, I am more interested in attending to one of its unexpected, geosonic transductions. The Golden Record was a direct inspiration for artist Trevor Paglen in his ambitious *Last Pictures* project (2012b), which involves inscribing specially curated images (also meant, in a way, to be representative of life on earth, like the audio material on the Golden Record) onto a metal disc that will orbit the earth aboard a telecommunications satellite designed, in principle, to maintain its geosynchronous orbit forever, long after human, or indeed any, life on earth will have ceased to exist. The images Paglen and his team selected emphasize technology, industrialization, exploitation of natural and animal resources, and large-scale military destruction, especially nuclear war. Unlike the sanguine mission of the Voyager, which is meant to go out into deep space and offer greetings to alien life forms from a vibrantly human planet, *The Last Pictures* is meant to orbit earth as a kind of tombstone for a humanity conceived of as already long-gone. On this temporal contradiction and bleak imagining of the future, Paglen has remarked:

> Some geologists point out that over the last hundred years or so, things like real estate markets have become geomorphic agents—fluctuations in commercial real estate markets can move more sediment than "natural" geomorphic processes like erosion or tectonics. Something like global warming is a perfect example of what I think of as a contradiction in time—global warming is an earth process that will play out over the next century or so, but it largely emerges from the industries controlled by business turnover times of a few months (at most). On top of that, we live in a political system where the turnover time of politics is a few years at a time (between elections, for example) and there's little incentive to address problems whose effects play out on a longer temporal scale than an election cycle. I think that global warming is a great example of how these contradictory time scales produce effects that humans have few credible means of dealing with.
>
> (2012a)

Bracketing the obvious existential threat of global warming, which, for Paglen, does not pose aesthetic problems per se, the real artistic dilemma of *The Last Pictures* is geological and temporal: how to inhabit multiple time scales simultaneously, and how/where to picture the present as both a catastrophe-in-progress and as an Ozymandias-like monument to an already-destroyed world, which future (nonhuman) viewers can look on and despair.

The end-of-the-world aesthetic resonates with Timothy Morton's concept of the hyperobject (discussed in Chapter 1), and I want to repeat the subtitle of Morton's 2013 book: Philosophy and Ecology *after the End of the World*. Morton and Paglen, *c.* 2012/13, are just beginning to catch up to Sun Ra who, by 1970, had composed a piece (with lyrics sung by June Tyson) titled "It's After the End of the World," the refrain of which plays at the beginning of Ra's 1974 film *Space is the Place*, with June Tyson chanting repeatedly, in a tone that's hard to describe: "It's after the end of the world, don't you know that yet?"[4]

The bizarre and provocative first image of *Last Pictures* is the reverse side of Paul Klee's *Angelus Novus* (1920), the drawing famously purchased by Walter Benjamin and immortalized as "the angel of history" in the latter's "Theses on the Philosophy of History" ([1940] 1968). The angel, in Benjamin's description, is gazing backwards into the past and being blasted forward into a future it cannot envision by the "storm [...] we call progress" (258). Like a good geological catastrophist, the angel, with "his face turned toward the past," does not "perceive a chain of events, [but rather] one single catastrophe which keeps piling wreckage upon wreckage and hurls it in front of his feet" (257). Between the imagery of the drawing and the pathos of Benjamin's language, the piece activates many key themes of Paglen's project: catastrophe within, or indeed *as* technological progress; ruination; a gaze tragically unable to envision the future, et cetera. Yet Paglen chooses not to present Klee's image itself, but instead the verso: a distinctly unspectacular display of administrative stamps and labels (in Hebrew and English) cataloguing the drawing in the collection of the Israel Museum in Jerusalem and indicating its temporary loan status for an exhibition at the Museum of Contemporary Art in Chicago. If the angel is blown backwards into an unknowable future, Paglen takes that metaphorical language literally and shows us what is on the other side of the image-as-material-object: not (just) a catastrophe-fueled future, but an archive, a museal infrastructure of preservation with its authenticating, proprietary protocols. This is Paglen's programmatic statement about art in the time-contradictory moment of global warming and planetary epitaphs: art is the archiving of the catastrophe that is happening *and* will have already happened, a minute and futile gesture of documentation that is meant not to represent, but merely to outlast the world of which it was a part.

Transduction-wise, I have just moved in this "Afterword" across a number of media, artefacts, and times: from the audio conceit of the Golden Record to Paglen's words on his *Last Pictures* project to Paul Klee's drawing to Walter Benjamin's geo/theological-catastrophist language about the drawing, and finally back to one of Paglen's selections, all to contextualize a visually boring image of museum bureaucracy. Let me take one more literal-minded step into this time-, media- and metaphor-bending loop. Benjamin emphasizes the gaze of the angel "looking as though he is about to move away from something he is fixedly contemplating. His eyes are staring, his mouth is open, his wings are spread" ([1940] 1968: 257). But the angel also has a prominent pair of ears which, in terms of the relative size and detail of their execution, are more dominant in the pictorial space than the eyes. Rich in visually inferred sound (and even richer via Benjamin's clamorous description), *Angelus Novus* is potentially as much an image of an inscrutable soundscape as of an inscrutable future, prompting viewers/listeners to speculate on the sounds the angel might already be hearing and raising an auditory/aesthetic question: how to translate the visual-historiographic problem of catastrophe-in-progress into sonic terms?

There is a scene at the beginning of Thomas Pynchon's 1973 novel *Gravity's Rainbow*, in which a British Second World War soldier, Capt. "Pirate" Prentice, contemplates the distant launch of a V-2 rocket through the vapour trails he catches sight of in a pre-dawn sky. Prentice is in London, and, knowing that London is the rocket's target, he briefly and morbidly imagines the rocket coming straight for him: "He takes some time lighting a cigarette. He won't hear the thing come in. It travels faster than the speed of sound. The first news you get of it is the blast. Then if you're still around, you hear the sound of it coming in" (8–9). Fortunately for Prentice, that particular rocket fizzles out mid-flight and falls out of harm's way off the English coast. This is just a passing moment in the novel, but from a geosonic perspective, it is a remarkable proposition that reconfigures the time-predicament of Klee's drawing. Sounds are usually attached to—and become part of—events that precede them, while taking on a reverberant afterlife as they propagate. In the case of the V-2 rocket, in which the military-industrial complex has managed to separate the speed of events from the sonic scale, the sonic anticipation of the blast resounds *after* it like an echo and could be the last sound you'll (never) hear if you happen to have been in its path. Pynchon proposes a calamitous, time-bended soundscape (which rapidly becomes an everyday reality in the novel) in which sounds give incipient clues to catastrophes that have already happened and remind listeners that they are, precariously and ambiguously, still alive. Pynchon's text (as I imagine it here, transduced through Klee's drawing, Benjamin's writing, Paglen's image, and the Voyager's Golden Record) intuits the paradoxical possibility of sound outside the flow of time, in ways analogous to Sun Ra practicing a contradictory geology beyond the logic of the earth. *Geosonics* listens across these multiple and contradictory temporal and planetary scales, attending to what is still unheard-of in a world-ending soundscape.

Acknowledgments

First thanks go to David Cecchetto, who supervised my doctoral work at York University, Canada, and who continues to be an invaluable interlocutor and advocate for my work. Janine Marchessault and Steve Bailey provided significant input and support for the dissertation, and many of the ideas for its individual chapters grew out of inspiring seminar work with Jody Berland, Greg Elmer, and Aleksandra Kaminska. Marc Couroux and Mack Hagood generously read the dissertation version of *Geosonics* and helped set me on the path to transform it into a book.

As a contract lecturer, I am grateful for colleagues at the University of Toronto, Canada, and beyond who recognize my research. I wish to thank Aleksandra Kaminska and Alanna Thain for selecting me as a postdoctoral fellow to collaborate on the Sociability of Sleep project at McGill University and the University of Montreal, Canada. Eleonora Diamanti and Alexandrine Boudreault-Fournier expressed support for my new work on sleep, which fed back into and enriched the sleep chapter in *Geosonics*. Paul Harris and Alexis Rider gave encouraging feedback on a version of the Sun Ra chapter. Scott Richmond has generously supported my role as the convener of a faculty/grad working group (on time and media) for the past two years at the University of Toronto's Centre for Culture & Technology. The rich and ongoing conversation of the Beating Time working group at the Centre influenced many of the ideas that went into the finishing stages of *Geosonics*. I wish to thank the members of this working group: Mitchell Akiyama, Kate Maddalena, Alexis Millares Thomson, and Anna Renken, as well as Brandon Hocura and Toleen Touq. Special thanks to JJJJJerome Ellis for sharing his time and creative work as part of that conversation.

My program director in Professional Writing & Communication, Lilia Topouzova, has been supportive not just of my effort to write a book, but to finish it. My colleague John Currie graciously stepped in at the last minute to take over one of my summer courses so I could work toward an important deadline on the manuscript. I thank Sarah Sharma, the director of the Institute of Communication, Culture, Information & Technology, who established a research fund for contract faculty, which has helped me to cover the publishing/licensing costs for this project.

I wish to thank the many students I have taught over the years, who continually confirm for me the value of academic dialogue and research. I have been greatly inspired by my students' creativity, hard work, and diversity of perspectives.

I want to acknowledge the academic organizations which have hosted conferences where I presented work that appears in *Geosonics*, and which have helped shape the sense of audience and intellectual landscape that I bring to my academic writing: the Music and Sound Studies Network of the German Studies Association (2014 and 2015), where kind and astute feedback from Cornell colleagues Samuel Frederick, Yuliya Komska, and Leslie Adelson proved decisive in my shift to writing about sound; the Society for Science, Literature & the Arts (2019 and 2021), which models interdisciplinarity and inventiveness; and Tuning Speculation (2016, 2017, and 2019), an intellectual community that very quickly became indispensable in terms of how I think and write about sound.

As someone who is not a trained musicologist or a jazz scholar, I have the unearned privilege in *Geosonics* of writing about Beethoven and Sun Ra. I want to thank the people I have played music with (teachers, friends, and players in formal and informal ensembles) and who indirectly empowered me to write about music: Martin Collin, Stacy Hobbs, Brian Trump, Peter Spaar, Sean Franzel, and Travis Workman; and Stephen Quinney, Kirk Shibley, and Paul McCulloch, and all the members of the Corktown Chamber Orchestra in downtown Toronto.

The two anonymous reviewers gave encouraging and insightful commentary which helped me to deliver a significantly stronger manuscript. For permission to reproduce Christine Sun Kim's artworks, I thank the artist herself, and acknowledge the quick and precise work of Gan Uyeda and Agathe Pinard in working out the details. For permission to reproduce Sun Ra's poem "The Empty Space," I want to acknowledge the lightning-quick and cooperative responses of Irwin Chusid and John Corbett. An earlier version of Chapter 3 appeared as "Toward an Aural Aesthetics of 24/7 Environments: Beethoven, Audio Stretching and Techno-Indeterminacy," *Journal of Aesthetics & Culture* 9.1 (2017). An earlier version of Chapter 6 appeared as "The Etymology of the World: Sun Ra, Geology, Poetry," *SubStance* 52.3 (2023).

My sincere gratitude to everyone at Bloomsbury who had a hand in the production of this book. Rachel Moore confidently oversaw the editorial process and generously supported the completion of the manuscript; Louise Dugdale made an excellent cover image; and RefineCatch helped in the final stages with thoughtful copy-editing.

Last thanks go to my family. I want to thank my parents Michael and Joyce Dittrich, and my sister Samantha Dittrich, for their encouragement during all my studies and academic work. My two sons, Daniel and Martin, help me think about music and sound all the time, and I can't imagine writing anything now without hearing what they think about it. Finally, this book would not exist without Stanka Radović. She read every word and discussed every idea; gifted me books about Sun Ra; shared desk space with me (eventually ceding it altogether!); and for years has taken precious time away from her own work to make mine possible.

Credits

Christine Sun Kim, *The Sound of Laziness*, 2016. Charcoal on paper, 19.5 x 25.5 inches (49.5 x 65 cm). Courtesy of the Artist and François Ghebaly Gallery, Los Angeles. Photo: Kell Yang Sammataro.

Christine Sun Kim, *The Sound of Apathy*, 2016. Charcoal on paper, 19.5 x 25.5 inches (49.5 x 65 cm). Courtesy of the Artist and François Ghebaly Gallery, Los Angeles. Photo: Kell Yang Sammataro.

Christine Sun Kim, *The Sound of Being Resigned*, 2016. Charcoal on paper, 19.5 x 25.5 inches (49.5 x 65 cm). Courtesy of the Artist and François Ghebaly Gallery, Los Angeles. Photo: Kell Yang Sammataro.

Christine Sun Kim, *Degrees of Deaf Rage within Educational Settings*, 2018. Charcoal and oil pastel on paper, 49.25 x 49.25 inches (125 x 125 cm). Courtesy of the Artist, François Ghebaly Gallery, Los Angeles, CA, and White Space, Beijing. Photo: Yang Hao.

Christine Sun Kim, *The Sound of Temperature Rising*, 2019. Acrylic on wall adapted from 2016 charcoal on paper drawing, Edition of 3, 1AP, 115.75 x 196.75 inches (294 x 499.74 cm). Installation view, PS120, 2019, Berlin, Germany. Courtesy of the Artist and François Ghebaly Gallery, Los Angeles. Photo: Stefan Korte.

Christine Sun Kim, *TBD, TBC, TBA*, 2015. Charcoal on paper, 11 x 15 inches (28 x 38 cm). Courtesy the Artist, White Space, Beijing, and François Ghebaly, Los Angeles. Photo: Yang Hao.

Sun Ra's poem "The Empty Space" appears courtesy of Sun Ra LLC and Corbett vs. Dempsey.

Notes

Introduction

1 See McPhee (1998: 126–7).

2 David Toop's chapter on Sun Ra from *Ocean of Sound* is titled, "If You Find Earth Boring . . ." ([1995] 2018).

3 For an excellent discussion of atmospheric sounding, communication, and media theory, see the chapter "Sounding" in Derek P. McCormack's *Atmospheric Things: On the Allure of Elemental Envelopment* (2018), 121–43.

4 In epistemological terms, this aspect of the soundscape is closely related to Steven Feld's concept of "acoustemology." See Feld (2015).

5 On the contradictions of Schafer's disavowal of technology and modern, audiophile listening practices, see Sterne (2013). On his conflation of the aesthetic with the moral in the underlying normative framework of the soundscape, see Thomspon (2017).

6 On the exclusion of Indigenous perspectives from the soundscape, see Robinson (2020). On how the soundscape resonates with an anti-urban, anti-immigrant settler-colonial nostalgia, see Akiyama (2019).

7 See Droumeva and Jordan's edited volume *Sound, Media, Ecology* (2019).

8 See Martin's piece, "Seeing the Sounds of 7th and Florida," featured on the website of her Black Sound Lab.

9 Jonathan Gilmurray makes a case for ecological sound art as a curatorial and conceptual term (2017). The volume *Environmental Sound Artists: In Their Own Words* contains artist statements and reflections from many key figures affiliated with the field (Bianchi and Manzo 2016).

10 "The history of deaf communication makes clear that sound is always already multimodal. Sound waves transfer between media (air, water, solids), and can be experienced by sensory domains beyond the ear" (Mills 2015: 52). See also Friedner and Helmreich's essay, "Sound Studies Meets Deaf Studies" (2012).

11 Deaf musician Evelyn Glennie assesses reverberant acoustic space of a concert hall by touch, that is, "how thick the air feels," suggesting that reverberation also has a tactile dimension that is available to deaf and hearing listeners ([2015] 2017). Glennie is a percussionist noted for performing barefoot, so that she can experience the reverberations of the sounds she makes, along with their resonance in the hall, through her feet.

Chapter 1

1 Aitken and Doug Aitken Workshop, "Sonic Pavilion," Vimeo [artist video based on the 2009 installation], available online: https://vimeo.com/152320997 (accessed April 26, 2020).
2 Cage in a 1965 interview, cited in Kahn (1999: 194).
3 Kahn (1999: 196).
4 Mixing the metaphor even further, the call of the earth (of the painting) is structured not unlike a phone call: even though Heidegger uses the term *Zuruf* (an acclamation), not *Anruf* (calling someone on the phone), the interpretation of the painting is framed within a distinctly modern (i.e., not ancient Greek) telephonic paradigm that recalls Avital Ronell's *The Telephone Book* (1989). I thank Sumanth Gopinath for the Heidegger/Ronell connection. It seems unlikely that Heidegger would have known of a device used by German trench forces in the First World War called an *Erdsprechgerät*, or earth speech device, a kind of wireless field telephone that worked by conducting voice signals through the earth. For more on such earth circuit systems in the history of telegraphy and radio, see Kahn (2013: 69–78).
5 See David Novak and Matt Sakakeeny on sound's "feedback loop of materiality and metaphor" (2015: 1); George Revill on the "'thingness' of sound as co-produced by the act or process of making, the materials which carry and transmit, and the means of receiving, sensing and making sense" (2016: 252–3); Frances Dyson on the fluctuations "between the sonic and the metaphoric, between sound as substance and sound as imagination" (2014: 54); David Cecchetto on sound "as a particular object that has no substance, as a kind of ideal object that nonetheless has real material effects (i.e., literal sounds)" (2013: 2); Mack Hagood on "account[ing] for sound in its remembered, imagined, phantom and linguistic forms" (2019: 27); and Robin James for a critique of how "sound embodies material immediacy and the metaphysics of a probabilistic universe" in neoliberal biopolitics (2019: 3–4).
6 Related to this "always again" structure, Hillel Schwarz makes an observation about the seventeenth-century origins of the "encore" in European music performance as an instance of the fundamental repetitiveness of sound's entry into human experience, memory, and culture. Audiences at that time, Schwarz claims, requested encores not at the end of a performance, but precisely in the middle, an interruption to the flow of the musical-dramatic time that paradoxically worked to complete the always-again listening process. The most moving or dazzling aspects of a musical performance could not be considered properly heard until they were heard *again* in the encore and only then (re)committed to memory and experience (2011: 24).
7 See Bennett (2010: 52–61).
8 For example, Alain Corbin's authoritative campanarian history of nineteenth-century France avoids references to soundscape or sonic environment, preferring

instead the notion of an "auditory landscape" ([1994] 1998). For him, the sonic is always to be contextualized within a sociocultural sphere, and the materiality of sound only matters insofar as it aids the reconstruction of a cultural history.

9 Kahn has discussed Oliveros' debt to occult, Theosophical traditions and her affinities to New Age and transhumanist ideas of technological enhancement (what Kahn calls a "spiritual technofuturism" [2013: 174–86]). The vexed points where the physics of music blurs into metaphysics and spirituality is a theme Kahn follows from Pythagoras to John Cage (Kahn 1999: 189–90; Kahn 2004).

10 This is the basis of Helmreich's "transductive ethnography," which has been productive for sound scholars like Mack Hagood (2019), who uses transductive ethnography to understand how physiology, audiology, neoliberal subjectivity, and changing ideas about sound, space, and agency coalescence around auditory phenomena like tinnitus, noise, and their technocultural remediations.

11 *Schnitt* means cut, but also implies combination. The word is tied to notions of shaping, arranging, and editing.

12 Zielinksi's theory of media interfaces as cuts aligns with Karen Barad's influential new materialist theory of agential realism, which depends on "agential cuts" that contour phenomena ontologically within the indeterminate relational flux of matter. What Barad calls "apparatus" in the context of scientific laboratory research aligns with what Helmreich or Zielinksi would call media, that is, instruments that produce temporary, shifting boundaries that enable material phenomena to become (partially and temporarily) knowable. Likewise, Barad's key relational concept, intra-action, describes, like transduction and interfacing, a mediating process that precedes and entangles ontology with materiality. See Barad (2007: especially chapter 4, 132–85).

13 Key texts on Deleuzian aesthetics and philosophy of the earth "as such" include: Grosz (2008), Woodard (2013), and Yusoff (2015).

Chapter 2

1 See Florian Dombois, "Auditory Seismology" (2011). There are some anecdotal exceptions. See Kathryn Miles (2017: 26–33). See also Kahn (2013: 133–7).

2 Scott Hessels' "Celestial Mechanics" (2005) is a complex data visualization piece adapted for viewing in planetariums. It doesn't display stars or planets, but rather documents flight paths, satellites, and non-pathed objects' flights (like helicopters and weather balloons) using data gathered from NASA. See also Jussi Parikka's discussion of satellite debris via Trevor Paglen's *The Last Pictures* project in *Geology of Media* (2015: 125–31).

3 In contrast to the aesthetic of Eno's music, Marina Peterson's ethnographic work in *Atmospheric Noise* (2021) shows how it is precisely the noise of airports and aerial traffic that implicates the sonic experience of atmosphere in the permeability of bodies and built environments. The atmosphere is only "up there" insofar as it is part of a "broader atmospheric sensorium" that entangles human bodies on the ground with atmospheric noise and pollution.

4 Side A of the record is a presentation of earthquake sounds, Side B of ionospheric "whistlers." Douglas Kahn discusses the latter at length in connection to the work of composer Alvin Lucier (Kahn 2013: 107–14), and mentions Side A in his chapter "Sound of the Underground" in connection both to the Cold War seismology of Speeth and the seismographic experimental compositions of Gordon Mumma, which I discuss later in this chapter. See Kahn (2013: 133–61).

5 Kahn discusses Benioff's interest in music, specifically his development of an electric cello and violin for which he built pick-ups using the same electronics as his industry-standard seismograph. According to Kahn, Benioff's instrument making grew out of what the geophysicist understood to be the deep analogy between the stress and strain of bowing a stringed instrument, and the seismic stress systems of earthquakes. The "seismographic fiddle" thus possessed "earth-scale acoustics" (Kahn 2013: 147–9).

6 Far be it from me to correct a pre-eminent geophysicist of the twentieth century, but the term "earth's acoustics," while rhetorically effective, is at best a mixed metaphor. The earth would be better said to have elastics, not acoustics.

7 Another three minutes and three seconds would bring this "silent" recording to a length of 4′33″. It is not clear if Benioff was familiar with John Cage's debut of *4′33″* in 1952, the year before *Out of This World* was released. It is also not clear if Cage knew a geophysicist had made a recording of inaudible, indeterminate, aleatory sounds within a year of the publication of his famous "silent" piece.

8 "'Earthquake for Home Use. It is understood as a condition of sale that Cook Laboratories, Inc., will in no way be responsible for damage this phonograph record may cause to equipment directly or indirectly. For users with wide-range woofers this disclaimer shall be construed to include neighbors as well, dishware and pottery'" (cited in Kahn 2013: 146).

9 Astonishingly, if somewhat morbidly, the poet imagines what kind of "primal sound" would emerge from tracing the coronal suture of the human skull. For Kittler, that "primal sound" would be the "real," the inhuman materiality that underlies the operation of all media, and that renders human subjectivity, and really all human social life, a kind of hallucinatory side effect of media technologies. This is not the place to go into Kittler's provocative and problematic media materialism, but it is necessary to remark that the sound that grounds a whole media ontology and history of audio technology is not exactly sound at all, but a transductive

fiction: an imagined sound that nobody ever actually heard reconstructed as a figure within a speculative literary text.

10 Kittler almost certainly has in mind Florian Dombois' audifications of the 1995 earthquake in Kobe, Japan, on the album *Earthquake Sounds* ([1986] 1999), when he makes the following remarks in the lecture, "Light and Series—Event and Thunder": "Take an earthquake like the one in Kobe [Japan] with thousands of casualties, seismographically record its inaudible slow vibrations, replay the signals of the entire horrific day in 10 seconds—and a sound will emerge. In the case of earthquakes that, like those in the Pacific, result from the clash of two tectonic plates, the sound will resemble a high-pitched slap, in the case of those that, like those in the Atlantic, are the result of the drifting apart of two continental plates, it will, conversely, sound like a soft sigh. Thus, the spectrum, that is, a frequency composition, gives the violent events timbre or quality: America becomes Asia. A short time ago I was privileged to hear the timbre of such quakes and I will not forget it for the rest of my life" (2006: 69).

11 Frantti and Leverault's paper is titled, "Auditory Discrimination of Seismic Signals from Earthquakes and Explosions" (1965). Kahn maps out these connections (2013: 155–6).

12 These are the pitches as I hear them on my piano, in any case. But I should disclose that I am primarily a double bassist, and geosonically, pitch discrimination, like quake/bomb discrimination, is not the forte of a bassist. Speeth initially wanted to train bass players specifically as test subjects in his research, "but he found that their pitch discrimination was not as keen as cellists'" and was "'pleased and somewhat amused to give employment to a bunch of cello players for a short period of time'" (Lauren and Christopher Speeth in correspondence with Douglas Kahn [2013: 151]).

13 Mumma's *Mographs* are a pre-cursor to another ambitious piano playback installation, namely Gordon Monahan's "A Piano Listening to Itself – Chopin Chord" (2010), an outdoor installation in Warsaw, Poland, in which a 24-hour loop of digitized Chopin recordings is mechanically fed into piano wires stretched from the tower of the Royal Castle to the soundboard of a piano in the square below. The music becomes audible as it vibrates the piano's soundboard, transforming the musical instrument into something that, in a sense, it already is: a transducer and a playback device. The lengthy piano wires also periodically transmit Aeolian vibrations when the wind conditions allow it, lending an aleatory, site-specific atmosphere to the otherwise closed playback circuit. See gordonmonahan.com; and also Mohanan (2010). The conceit of a 24-hour cycling sound installation that (potentially) opens itself onto a geosonic time-scale is a problem I discuss in Chapter 3.

14 The *Large Size Mograph* recording time does not derive from Mumma's four-hands collaboration with Tudor in the 1960s, but from *Gordon Mumma: Music for Solo Piano 1960–2001*, performed by Daan Vandewalle, New World Records, 2008.

15 It is worth noting, as another curious coincidence of ear-drums and earth quakes, that one of the most influential digital audio tools in contemporary music, Auto-Tune, was invented by a former seismologist, Dr. Andy Hildebrand, who was recognized with the Technical Grammy in 2023 for his achievement. When musicians use Auto-Tune, it's fair to say they are being creative to a fault. Thanks to Marc Couroux for this reference.

16 See Kahn (2013: 115–21). See also the 2008 DVD with documentary footage, interviews, and audio track of one of the performances.

17 See also my discussion of Aitken's "Sonic Pavilion" in Chapter 1.

18 *Inner Earth, A Seismosonic Symphony* (1999) by Wolfgang Loos and Frank Scherbaum (a composer and seismologist, respectively) uses digital techniques to create a five-movement ambient work out of seismographic data, which the back cover of the CD describes as "new musical territory beyond chill-out music and Musique Concrète." Mark Bain's "Live Room: Transducing Resonant Architecture" (1998) uses seismographs to playback mechanically induced "seismicity" in built environments, a project that was initially realized in a former Cold War instrument research facility at MIT. The "Seismodome," a project tied to Columbia University's Geology Department, retrofits a cutting-edge data aggregation tool called "Instaseis" (a kind of Instagram for earthquakes) for audiovisual simulations of earthquakes in immersive-surround presentation in planetariums.

19 User comments on YouTube cast doubt on the viability of such a small sensor/implant to contain a Bluetooth, a power source, and to convey the complex seismographic data with any nuance or texture. I myself lack the technical or engineering know-how to weigh in with any authority, but it does seem unlikely that the complex information played back by audio seismograms could be meaningfully transferred to a buzzing or pinching sensation felt by the elbow.

Chapter 3

1 Although Douglas Kahn (1997), among others, is critical of the extent of Cage's departure from the privileged aesthetics of the concert hall. See also my remarks on Cage and microphones in Chapter 1.

2 Indeterminacy is thus linked to microphony. The score of *0′00″* specifies that performers should "perform any disciplined action" that is part of their everyday life on surfaces fitted with contact microphones. At the debut, Cage conducted the business of his personal correspondence, opening, reading, and responding to letters while seated at a desk (with microphones) on stage (Kahn 1999). The redirection of aesthetic attention (from musical "objects" to daily processes) is facilitated by

microphones, but the aesthetics of the piece insist on the inherent musicality of processes that only become so when transduced microphonically.

3 Evens' opening chapter is a tour de force that links psychoacoustics, musical performance, and recording technology to an all-encompassing sonic materialism enacted through the contraction of hearing.

4 In this sense, *9 Beet Stretch* is not dissimilar from Douglas Gordon's *24 Hour Psycho* (1993)—another work premised on 24-hour remediation of a canonical work—in that the defamiliarization of the altered playback speed is matched, frame by frame, grain by grain, with the formal refamiliarization of a Hitchcock's masterpiece. I thank Steve Bailey for confirming for me that Gordon's installation has no audio component.

5 It is worth noting that Beethoven himself experimented with compositional conventions to show how musical form itself can be "stretched." For example, the second movements of op. 57 and op. 111 (piano sonatas no. 23 "Appassionata" and 32, respectively) as well as the third movement of the String Quartet in A minor (op. 132) play with duration, variation and modulation in ways that anticipate the effects of audio-stretching. I am grateful to an anonymous reviewer of an earlier version of this chapter (published in *Aesthetics & Culture* [Dittrich 2017]) for these references.

6 For a lucid critique of Hansen's positing of the primordiality of the body with respect to technology and language, see Cecchetto (2011).

7 In a separate chapter, Hansen does discuss Gordon's *24 Hour Psycho*, but his focus is the work's foregrounding of viewer anticipation in the slowness of its playback. He mentions in passing that the screening of the film would have to be limited "to the opening hours of a museum or gallery" and that "no perception of the whole film is possible" (2004: 244). In the digital light years of cultural time that have elapsed since the 2004 publication of Hansen's book, 24-hour works like Christian Marclay's *The Clock* (2010) regularly play in full duration at major galleries and venues, not to mention the *9 Beet Stretch* itself in its "live" installation and portable formats. A more recent 24-hour music video, Pharrell Williams' *24 Hours of Happy* (directed by Yoann Lemonie and We Are From LA) has been reviewed, in its full duration by a number of bloggers. Both Marclay's and Williams' pieces (the latter via its continuous playback on its website) are synced to the actual clock time, whereas the online incarnations of Inge's piece are synched to the original start time of the debut of Beethoven's symphony in Vienna at 7 p.m. (on May 7, 1824). In any case, further on in the chapter I address the gap between Hansen's putative impossibility of a 24-hour duration *c.* 2004 and the seeming plausibility of devoting 24 hours (continuously, or in digitally separated chunks) just over a decade later.

8 For an insightful review of Crary's book that both appreciates its critical gesture, but also draws attention to some of the overgeneralizations that such a broad argument employs, see Parry-Davies (2016).

Chapter 4

1 It is important to note that "deafness" (lower case d) refers to a range of physiological states or conditions, while "Deaf" (capital D) refers to an actively assumed cultural identity. In this chapter, I will follow Kim's practice of calling herself a Deaf artist. But I use "deaf" more often in my discussion since it is more at the level of sensory and communicative access (and not necessarily at a Deaf identity politics), where I would situate the central intervention of Kim's work. See also Friedner and Helmreich (2012: 74) and Davidson (2021: 226–7 n. 10) on the usage of deaf/Deaf in media and disability studies contexts, respectively.

2 Audism refers to the systemic marginalization and stigmatization of deaf culture and experience, a version of ableism specific to the ability of hearing. Oralism, used later in the chapter, refers to the systematic teaching of auditory speech and lip-reading (to the exclusion of sign language) to deaf people, and is a historically specific form of audism which, taken as large-scale linguistic and cultural suppression, has been likened by Deaf scholars to a form of colonialism. For a concise summary of the substantial literature in Deaf studies, see Tan (2020: 241–2).

3 Harbisson shares biographical and personal anecdotal material in a number of print interviews with journalists as well as promotional talks and videos (usually with corporate sponsors like Samsung, Vodaphone, Google, etc.) available online. His 2012 TED Talk "I Listen to Colour" and his 2016 talk at Google "What's It Like to be a Cyborg?" (co-presented with Moon Ribas) are his most concise and extensive (respectively) self-presentations. Unless otherwise indicated, all the information I offer on Harbisson's life, art, and implant is drawn from these two sources.

4 Available online: https://www.cyborgfoundation.com.

5 Available online: https://www.transpeciessociety.com.

6 This history starts in European modernism of the 1920s and 1930s as a cultural ambivalence toward industrialization, urbanization, mass media, and world war (Biro 2009). Technology was seen as both a destructive and transformative force, and cultural critics have pointed out that representations of the (predominantly masculine) body in modernist literature and art were both haunted "by the spectre of the damaged body of the worker-soldier" (Foster 2004: 114) as well as "characterized by a desire to *intervene* in the body, to render it modern by technologies, which may be biological, mechanical, or behavioural (Armstrong 1998: 6). A proto-cyborgian prosthetic logic obtained, in which organic bodies were seen as both fragile and incomplete, but also monstrously improvable by technology. This modernist/masculinist take on the cyborg predates but informs the classic definition of the term, which refers to the cybernetic organism circulating in artificial environments of space travel (Clines and Kline 1960). The postmodern version of the cyborg is synonymous with Donna Harraway's reinvention of the

term as a category-defying entity that rewrites modernist techno-ambivalence within the control logic of information societies. For Haraway, cyborgs both express information's desire for technological control over bodies, but at the same time embody an unstable blend of identities and positions that harbor the potential for a critical, feminist, and speculative techno-politics (1990). A generation of performance artists (some discussed in the second part of this chapter, like Orlan, Eduardo Kac, and Stelarc) develop Haraway's cyborg aesthetic into critical art practices that thematize the body as a modifiable interface of technical systems. By contrast, Harbisson's cyborgism returns to the prostheticism of the 1920s, but lacks the sense of critical potential and of subversive, uncontainable social energy with which the cyborg has been invested by a previous generation of artists and theorists.

7 Such interpretations tend toward the sensationalistic: an internet search of "transpecies" might yield, besides Harbisson's own website, links to tabloid stories of people who identify as cats, dragons, or elves and spend thousands of dollars on surgery, implants, tattooing, and piercing to modify their bodies accordingly.

8 Mark O'Connell's *To Be a Machine* (2017) is a lucid and critical survey of the major figures, factions and paradoxes in contemporary transhumanism. In terms of the scope of the movement, O'Connell shows that both Oxford and MIT have transhumanist research institutes; robotics and AI research overlap with transhumanism in a number of professional and institutional contexts; Silicon Valley is open for transhumanist business, symbolized most prominently in Ray Kurzweil (popularizer of "technological singularity") as director of engineering for Google; DARPA and military R&D can be linked to transhumanism as much as DIY "grinders" who make their own experimental body modifications in basement lab spaces across North America; and there are even Mormon transhumanists, and other more ecumenical, explicitly religious variants of transhumanism. In O'Connell's book, the movement emerges as a strange fusion of speculative theology, self-help, American "manifest destiny," risk theory, and technocapitalism. Another useful introductory text (though without O'Connell's outsider scepticism) is Max More and Natasha Vita-More (eds.), *The Transhumanist Reader* (2013), which uses an academic format to present a range of promotional and programmatic transhumanist texts.

9 The editorial introduction to a recent volume on *The Matter of Disability* offers an unambiguous critique of the transhumanist take on disability: "Transhumanism effectively extends the most dangerous inclinations within humanism in that the proponents invest in the capacity of a human-directed escape from disability and other late eugenical dreams of an exceptionally capacitated humanity beyond our current one" (Mitchell, Antebi, and Snyder 2019: 4). Gregor Wolbring critiques the ableism of transhumanism along similar lines (2006).

10 Such tokenization approaches the notion of "narrative prosthesis," namely, the ways that figures of disability have both a grounding and mystifying role in literary texts (Mitchell and Snyder 2001). What Mitchell and Snyder develop as a narrative and literary-historical concept can be, arguably, applied to political rhetoric as well.

11 The "Transhumanist Declaration" (2012) mentions overcoming humanity's involuntary "confinement to planet Earth" (More and Vita-More 2013: 54) and More's "Philosophy of Transhumanism" claims that transhumanists "look to expand the range of possible future environments for post-human life, including space colonization and the creation of rich virtual worlds" (4).

12 See, among many discussions, three edited volumes: Mike Featherstone (ed.), *Body Modification* (2000); Joanna Zylinska (ed.), *The Cyborg Experiments* (2002); and Arthur Kroker and Marilouise Kroker (eds.), *Critical Digital Studies* (2008), 44–512.

13 One leading narrative of the posthuman is traced by Katherine Hayles in *How We Became Posthuman* (1999), describing how the history of cybernetics and information theory is implicated in a redefinition of the human body as a "posthuman" cyborg, that is, a technically mediated ensemble of embodiment and virtuality. Similar ground, with an emphasis on the future of distributed cognition rather than its historical evolution, is covered with a more sanguine touch by Andy Clark in *Natural-Born Cyborgs* (2003). A philosophically robust theory of originary technics (and by extension, a kind of originary cyborgness of the human) has been developed out of Heidegger's work on technology and Derrida's notions of supplementarity and prosthesis by thinkers such as Bernhard Stiegler and Mark Hansen. Judith Butler (1993) and David Wills (1995) offer similar accounts of the body's incomplete origins, emphasizing language and discourse, rather than technology, as the body's originary and continual supplement.

14 See the Wikipedia article on "Sonochromatism," which mainly cites Spanish media articles about Harbisson.

15 See Peter Pesic, *Music and the Making of Modern Science* (2014), 121–31.

16 In a presentation of an earlier version of this chapter at the "Tuning Speculation V Conference" (2017), many participants (with professional backgrounds in electronic and digital sound production) expressed doubt that such a prosthetic could actually work in the way that Harbisson describes. Of course, whether the prosthesis actually "works" or is just a mere rhetorical performance is all the same from the perspective of a transductive listening analysis.

17 Hansen, *Bodies in Code* (2006).

18 This problem (in ancient to modern philosophy and music) is treated by Daniel Heller-Roazen, *The Fifth Hammer* (2011).

19 "The ultimate private acoustic space is produced with headphone listening, for messages received on earphones are always private property" (Schafer 1997: 118). As a visible marker of an audio secret, the antenna functions like a Walkman in the

classic essay, "The Walkman Effect" by Shuhei Hosokawa, where theatricality and spectacle supplant the matter and technique of listening (2012).

20 Kim, "Courtier as Courier" (2013).

Chapter 5

1 In this paragraph, I paraphrase freely from Steven W. Lockley's and Russell G. Foster's *Sleep* (2012), which deftly synthesizes much of the recent science on sleep, along with the history of its interdisciplinary study in medicine, neuroscience, and sociology.

2 Decades of social constructionist approaches have effectively done away with stable, self-evident, and essentialist notions of the body. For an overview, see Bryan S. Turner (2012) and Lisa Blackman (2008).

3 Decisive experiments on the gap between internal and externally determined circadian rhythms involved researchers and subjects descending into perpetually dark caves (such as Nathaniel Kleitman in Mammoth Cave, Kentucky, in the 1930s) or designing sealed, windowless bunkers where subjects could only use artificial light sources (as in Jürgen Aschoff's and Rütger Wever's pioneering experiments in chronobiology in the 1960s). See Benjamin Reiss (2017).

4 John Durham Peters makes a different, but not unrelated point about lists in *The Marvelous Clouds*, where lists "repeatedly point to how the world escapes our concepts" and constitute a "crisis of uncontainable relevance" for media theory" (2015: 9). In the age of Google, lists represent for Peters the sometimes exhilarating, sometimes exhausting trade-off between stable forms of knowledge and endlessly proliferating knowability. For Nancy, the potential exhaustion of the list appears as the juxtaposition of bodies and the sliding of scales in the (in)distinction of sleep.

5 Steve Goodman's *Sonic Warfare* (2012) focuses on the (counter-)weaponization of sound in a Deleuzian politics of the war machine. Annahid Kassabian's *Ubiquitous Listening* (2013b) explores how the circulation of affect in music demands a rethinking of subjectivity and attention in listening. And the volume edited by Marie Thompson and Ian Biddle, *Sound, Music, Affect* (2013), listens for the ways in which music in particular facilitates and distributes affect within sonic cultures.

6 See, for example, the volume edited by Karolina Doughty, Michelle Duffy, and Theresa Harada, *Sounding Places* (2019), as well as essays by George Revill, "How is Space Made in Sound?" (2016), and Michael Gallagher, "Sound as Affect" (2016).

7 "Identity is a position left behind by the work of affect and while it has been perceived as positional—that is, as static—it now looks like a constant process. [...]

Affect both conditions and enacts identities and identifications, and it does so not within bodies, but across them" (Kassabian 2013b: xxix).

8 Some musicians even find this kind of nonconscious listening crucial for creation and improvisation. John Corbett's excellent book, *A Listener's Guide to Free Improvisation* (2016) recounts the following anecdote: "[B]assist Barry Guy once told me he believed in hearing through his bones. Because sound travels as much as four times as fast in liquids and non-porous solids than in air, he insisted that he could sense the music immediately, long before it made its way through the convoluted path of the outer ear, drum, stirrups, cochlea, basilar membrane, organ of Corti, sensory hairs, auditory nerve, to the brain" (110). Such a hearing is perhaps materially possible, though it is practically or experientially only imaginary, which does nothing to undermine its status as listening.

9 See Hayles' chapter 3, "The Cognitive Nonconscious and the New Materialisms" (2017: 65–85).

10 For Malabou, any study or theory of the brain implies necessarily a theory of power, control, or governance on some level: "There is today an exact correlation between descriptions of brain functioning and the political understanding of commanding" (2008: 32).

11 Such discourses have been addressed in a critical-historical context by: Anson Rabinbach, *The Human Motor* (1990); and Alain Ehrenberg, *The Weariness of the Self* (2010).

12 See Joseph Lanza's ironically laudatory account in *Elevator Music* (2004). Goodman's take on Muzak is more cautionary, see *Sonic Warfare* (2012: 141–8).

13 It is worth noting that Rabinbach's exhaustive (no pun intended) cultural history of fatigue in the nineteenth and early twentieth centuries concludes with his assertion that after the Second World War, the technoscientific paradigm of the productivist human body (including techniques to enhance productivity and minimize fatigue) came to an end. It is likely that the science of sleep (just gaining legitimacy in the postwar era) as well as the work-enhancing powers of Muzak stepped in to fill the void left by the long nineteenth-century's materialist, positivist conception of the body at work.

14 On 24/7 programming, see Lanza (2004: 45). On the use of Muzak in agribusiness, Lanza cites a 1973 *Rolling Stone* article: "'There was a situation when the National Stockyards in Illinois had too many 'dark cutters,' which happens when the release of adrenalin makes the blood congeal and the meat turn. They put the Muzak in and it calmed the cattle as they went to the hereafter'" (152).

15 I further explore sonic sleep media, or what I call "so(m)niferous media," and the contradiction of auditory-attention-as-sleeping in two essays, "Listening at Night" (2022) and "Bedtimes Stories" (2023a).

16 Speaking specifically of Western "classical" music, Kassabian has pointed out that, in the spaces of ubiquitous listening, much of what used to be considered "foreground" has been relegated to the background.
17 Hennesey (2016).
18 Brown (2018).
19 Consider by contrast August Browne's comment in his review of *Sleep*: "I would have loved to talk with fans about how 'Sleep' worked on them. But it seemed not just rude to go around waking people up to chat—it would have been inauthentic to Richter's goal for 'Sleep.' This was music as a physically communal, but profoundly interior and unknowable experience. The meaning is made when the mind drifts elsewhere" (2018).
20 For an eloquent take on rest as a practice of resistance within an African American cultural and religious framework, see Tricia Hersey's *Rest is Resistance* (2022). My contextualization of Ouzounian's and the Obadike's works is based largely on their own commentary as presented in talks at the Sleep Salons of the Montreal-based research project Sociability of Sleep (2022).

Chapter 6

1 John Szwed's *Space is the Place* ([1997] 2020) is a magisterial biography, which is based on archival research, musical and literary analysis, and interviews with Ra's band members, but also draws freely on interviews conducted by the author with Ra himself. Szwed's "bioethnography" (xx) thus weaves together the stories that circulate about Ra with Ra's own (often) self-mythologizing voice, giving the reader a clear historical frame and ample imaginative space to engage with Ra's story and work in a multiplicity of different ways. Throughout this chapter, I refer to Szwed's book both for its critical perspective and for its citing and paraphrasing of Ra's own language, as seen in this quote on "birth/earth." Ra revisits the same wordplay in the 1972/1980 poems titled, "Be-earthed" (Sun Ra 2005: 175).
2 Throughout his life, Ra obfuscated or concealed the circumstances of his actual birth (as Herman Poole Blount, on May 22, 1914, in Birmingham, Alabama). Ra maintained he was not born at all, but came from another planet or another time. Many of Ra's biographers, when faced with the obligatory task of discussing their subject's birth and death, will rather refer (as Ra himself did) to his arrival on and departure from planet earth. See Szwed ([1997] 2020: 4–5, 380); Paul Youngquist, *A Pure Solar World* (2016: 7, 260–3); Lock (1999: 47, 51); and Campbell, Trent, and Pruter (2021).
3 See the concept and practice of "Black Vernacular Phenomena" in Havis (2023).

4 Ra does not use the term "Afrofuturism," but many of the artists and critics who do (in the wake of Mark Dery's coining it in 1994, a year after Ra's death) identify Ra as Afrofuturism's originator. Afrofuturism refers to cultural and artistic production in the African diaspora that fuses folk culture, technology, and speculative mythologies of the past and the future. Szwed characterizes Ra's Afrofuturism as a kind of "retrofuturism, the return to the past to complete unfinished projects and to bring forward neglected or suppressed culture" ([1997] 2020: xxv). For detailed, biographical accounts of the impact of race, racism, and African American identity on Ra's proto-Afrofuturist views on history, myth, space, and music, see Lock (1999), Youngquist (2016), and Sites (2020). Alondra Nelson offers a classic theoretical and critical exposition of Afrofuturism as a kind of foil to cyberpunk discourse in, "Introduction: Future Texts" (2002).

5 In the essay, "The Air Spiritual Man," Ra uses the preposition "on" as adjective and a verb to emphasize the movement that is inherent to the seemingly static nature of being-on: "There are [*sic*] a special category of words that move in spirals. These are the *on* words that *on* things. The words of movement . . ." (emphasis added). See also Ra's insights "on / the / outer reach / of the unending / ON / ONNESS 'O N.'" in Ra's 1972 poem, "The Pivoting Planes," an even more cryptic and compressed version of Ra's verbo-sonic cosmogony. You can hear Ra reciting the poem—with heavy reverb and swirling background noise—in a radio broadcast/recording of his poetry in 1976, Sun Ra, "Sun Ra Reads Poetry 12/25/1976 WXPN Philadelphia," available online: https://www.youtube.com/watch?v=TmdqjXTkrD8 (accessed June 20, 2023).

6 See *The Wisdom of Sun Ra: Sun Ra's Polemical Broadsheets and Streetcorner Leaflets* (2006), compiled and introduced by John Corbett. The writings reflect Ra's research into Black identity as part of the Thmei Research group, a Black intellectual collective founded by Ra and his friend and partner Alton Abraham in the 1950s. Although the documents were collectively authored, I follow many biographers in attributing the dominant voice behind them to Ra himself. My classification of these documents as "writings" is ambiguous: although they were typed up, mimeographed, and distributed as leaflets, the writings also circulated as aural texts delivered by Ra himself, as part of the lively Black oral culture (including street-corner preaching and religious and cultural debate) of Chicago's Washington Park milieu.

7 See Sites, *Sun Ra's Chicago* (2020), especially the chapter "'Sound so Loud It Will Wake the Dead.'" My discussion relies on Sites's astute commentary and contextualization of these broadsheets in the currents of mid-century counter-cultural Black thought.

8 On the "creative toponymy" of the broadsheets (their projecting of ancient times and place-names onto the social landscape of 1950s Chicago), see Sites (2020: 112–15).

9 See the broadsheet "jesus said, 'let the negro bury the negro'" (66–7). Youngquist calls this a "bogus but provocative etymology" (2016: 58).

10 Jayna Brown links the equation of nothing and blackness in Ra's work to his exploration of provocative and emancipatory possibilities of being: "*Nothing*, the subject of many of Ra's poems, is blackness. But nothing is not the negation of being itself. Blackness is nothing—that is, anterior to ontological possibility, but only on earthly terms. *Nothing* is the 'freedom not to be,' the refusal to respect sovereignty or acquiesce to its terms" (Brown 2021: 174).

11 "My Music Is Words" was originally published in the journal *The Cricket: Black Music in Evolution*, edited by Amiri Baraka (then Leroi Jones), a collaborator and admirer of Ra. The journal advocated the Ra-like idea of a continuum between Black poetry and music, as well as their synergistic, liberatory cultural power. For more on the ways Ra's poetry fits (but also misfits) in Black radical poetry of the 1960s, see Edwards' chapter, "The Race for Space: Sun Ra's Poetry," in *Epistrophies* (2017: 120–53).

12 Robert Campbell and Christopher Trent have published an authoritative discography, *The Earthly Recordings of Sun Ra*. The first edition (1994) was 247 pages long and listed roughly 520 sessions. The second edition (2000) listed 789 sessions in its 847 pages. The third edition is in progress.

13 See *When Angels Speak of Love* (1966, recorded 1963) and *Art Forms of Dimensions Tomorrow* (1965 and 1972, recorded 1961–2).

14 In the notes for the 1992 rerelease of the record on Evidence Records, Michael Shore calls the album "proto-psychedelia," saying that when the hippies invented "musical mind expansion" on the West Coast, it was actually "Ra [who] had been there first." Irwin Chusid also notes that "Cosmic Tones prefigures by a few years the outer dimensions of psychedelia" in the 2014 reissue on Enterplanetary Concepts. Rodger Coleman echoes this proto-psychedelic reading (2022: 43).

15 My analysis here resonates with what Matthew Omelsky identifies as the formal "accumulations" and "spacings" across Ra's poetry and music as vectors of Ra's "outward-oriented fugitive time consciousness" (2023: 161).

16 This text was later published as a poem titled "To The Peoples Of Earth" in *The Immeasurable Equation* (Sun Ra 2005: 390).

Afterword

1 In interviews and writings, Ra consistently dismissed sleep as a waste of creativity and a purgatory-like state between life and death. Numerous observations from fellow musicians and anecdotes from journalists attest to Ra's lifelong insomnia. He

might fall asleep briefly at the piano during rehearsal, or in the middle of an interview with a journalist, but when he woke back up, he would say he had been "cogitating," not sleeping (Szwed [1997] 2020: 96).

2 See the opening set piece in my essay, "The Etymology of the World," for further reflections on etymology and/as geomorphology (Dittrich 2023b).

3 Andrew Knoll's chapter, "The Catastrophic Earth," provides a concise overview of geological catastrophism for nonexperts (2021: 171–93).

4 In Black and Caribbean studies, the notion of (life after) the end of the world is linked to the transatlantic slave trade, and the seminal theoretical text in this regard is Édouard Glissant's *Poetics of Relation* ([1990] 1997). Kara Keeling cites Ra's "It's After the End of the World" song to evoke the "apocalyptic catastrophe" of the transatlantic slave trade and the significance, for a range of speculative Black and Afrofuturist artists and thinkers, of conceiving of the future from the perspective of an already-ended world (2019: 54). In a polemic against what she calls "White Geology," Kathryn Yusoff does not mince her words: "The Anthropocene might seem to offer a dystopic future that laments the end of the world, but imperialism and ongoing (settler) colonialism have been ending worlds for as long as they have been in existence" (2018: xiii). The after-world motif opens the possibility of further exploring catastrophic time across geology, Black studies, and environmental media studies.

Bibliography

Adorno, Theodor W. [1949] 2019. *Philosophy of New Music*. Ed. and trans. Robert Hullot-Kentor. Minneapolis, MN: University of Minnesota Press.

Adorno, Theodor. [1951] 2005. *Minima Moralia: Reflections from Damaged Life*. Trans. by E. F. N. Jephcott. London: Verso.

Adorno, Theodor W. 2002. *Essays on Music*. Introduction and Commentary by Richard Leppert, trans. Susan H. Gillepsie. Berkeley, CA: University of California Press.

Adorno, Theodor W. and Max Horkheimer. 1972. *Dialectic of Enlightenment*. Trans. John Cumming. New York: Continuum.

Aitken, Douglas and Doug Aitken Workshop. "Sonic Pavilion." Vimeo [artist video based on the 2009 installation]. Available online: https://vimeo.com/152320997 (accessed April 26, 2020).

Akiyama, Mitchell. 2019. "Nothing Connects Us but Imagined Sound." In Milena Droumeva and Randolf Jordan (eds.), *Sound, Media, Ecology*, 113–30. Cham: Palgrave MacMillan.

Alcántara, Carmela, Luciana Andrea Giorgio Cosenzo, Weijia Fan, David Matthew Doyle, and Jonathan A. Shaffer. 2016. "Anxiety Sensitivity and Racial Differences in Sleep Duration: Results from a National Survey of Adults with Cardiovascular Disease." *Journal of Anxiety Disorders* 48 (May): 102–8.

Alcántara, Carmela, Sanjay R. Patel, Mercedes Carnethon, Sheila F. Castañeda, Carmen R. Isasi, Sonia Davis, Alberto R. Ramos, Elva Arredondo, Susan Redline, Phyllis C. Zee, and Linda C. Gallo. 2017. "Stress and Sleep: Results from the Hispanic Community Health Study/Study of Latinos Sociocultural Ancillary Study." *SSM Population Health* (December 3): 713–21.

Armstrong, Tim. 1998. *Modernism, Technology and the Body: A Cultural Study*. Cambridge: Cambridge University Press.

Augoyard, Jean-François and Henry Torgue, eds. 2005. *Sonic Experience: A Guide to Everyday Sounds*. Trans. Andra McCartney and David Paquette. Montreal and Kingston: McGill-Queen's University Press.

Bachelard, Gaston. 1994. *The Poetics of Space*. Trans. Maria Jolas. Boston, MA: Beacon Press.

Barad, Karen. 2007. *Meeting the Universe Halfway: Quantum Physics and the Entanglement of Matter and Meaning*. Durham, NC: DukeUniversity Press.

Beethoven, Ludwig van. 1997. *Symphony No. 9 in D minor, Op. 125: "Choral."* Mineola, NY: Dover Publications.

Benioff, Hugo and M. G. Morgan. 1953. [LP] *Out of This World*. Cook Recordings. Accessed via youtube.com. "Earthquake Recording from Emory Cook

Labs—WHEN AUDIOPHILES WERE INSANE!" Uploaded by Mikegoat, March 11, 2014. Available online: https://www.youtube.com/watch?v=D9pLEmdbK5A&feature=emb_logo (accessed February 28, 2020).

Benjamin, Walter. [1940] 1968. "Theses on the Philosophy of History" in *Illuminations*. Ed. with an introduction by Hannah Arendt, trans. Harry Zohn, 253–64. New York: Schocken Books.

Bennett, Jane. 2010. *Vibrant Matter: A Political Ecology of Things*. Durham, NC: Duke University Press.

Bianchi, Frederick and V. J. Manzo, eds. 2016. *Environmental Sound Artists: In Their Own Words*. Oxford: Oxford University Press.

Biro, Matthew. 2009. *The Dada Cyborg: Visions of the New Human in Weimar Berlin*. Minneapolis, MN: University of Minnesota Press.

Bjornerud, Marcia. 2018. *Timefulness: How Thinking Like a Geologist Can Help Save the World*. Princeton, NJ: Princeton University Press.

Blackman, Lisa. 2008. *The Body: The Key Concepts*. Oxford: Berg.

Brown, August. 2018. "The All-night, Outdoor Concert 'Sleep' Creates a Calming Reprieve with a Sense of Loss." *Los Angeles Times*, July 29.

Brown, Jayna. 2021. *Black Utopias: Speculative Life and the Music of Other Worlds*. Durham, NC: Duke University Press.

Butler, Judith. 1993. *Bodies That Matter: On the Discursive Limits of "Sex."* New York: Routledge.

Cage, Jonathan. [1937] 2011. "Future of Music: Credo." In Caleb Kelly (ed.), *Sound: Documents of Contemporary Art*, 23–5. London: Whitechapel.

Cage, Jonathan. [1966] 2008. [DVD] *Variations VII*. New York: Experiments in Arts and Technology. Artpix. 41 min (film), 85 min (audio).

Cage, Jonathan. 2011. *Silence: Lectures and Writings. 50th Anniversary Edition*. Middleton, CT: Wesleyan University Press.

Campbell, Robert L. and Christopher Trent. 2000. *The Earthly Recordings of Sun Ra*, 2nd ed. Redwood, NY: Cadence Jazz Books.

Campbell, Robert L., Christopher Trent, and Robert Pruter. 2021. "From Sonny Blount to Sun Ra: The Chicago Years." Revised December 12, 2021. Available online: https://campber.people.clemson.edu/sunra.html (accessed June 20, 2023).

Carruth, Allison. 2014. "The Digital Cloud and the Micropolitics of Energy." *Public Culture* 26 (2): 339–64.

Castiglione, Baldesar. [1528] 1902. *The Book of the Courtier*. Trans. Leonard Eckstein Opdycke. London: Duckworth & Company. E-book: April 2022, Project Gutenberg. Available online: https://www.gutenberg.org/files/67799/67799-h/67799-h.htm#r211 (accessed December 12, 2023).

Cecchetto, David. 2011. "Deconstructing Affect: Posthumanism and Mark Hansen's Media Theory." *Theory, Culture & Society* 28 (3): 3–33.

Cecchetto, David. 2013. *Humanesis: Sound and Technological Posthumanism*. Minneapolis, MN: University of Minnesota Press.

Chua, Daniel K. L. and Alexander Rehding. 2021. *Alien Listening: Voyager's Golden Record and Music From Earth*. New York: Zone Books.

Clark, Andy. 2003. *Natural Born Cyborgs: Minds, Technologies and the Future of Human Intelligence*. Oxford: Oxford University Press.

Clines, Manfred E. and Nathan S. Kline. 1960. "Cyborgs and Space." *Astronautics* 26 (7): 74–6.

Coleman, Rodger. 2022. *Sun Ra Sundays*. Edi. Sam Byrd. Grand Forks, ND: The Digital Press at the University of North Dakota.

Connor, Steven. 2006. "Strings in Earth and Air." Lecture given at the Music and Postmodern Cultural Theory Conference, Melbourne, December 5.

Connor, Steven. [2005] 2011. "Ears Have Walls: On Hearing Art." In Caleb Kelly (ed.), *Sound: Documents of Contemporary Art*, 129–39. London: Whitechapel.

Corbett, John. 2016. *A Listener's Guide to Free Improvisation*. Chicago, IL: University of Chicago Press.

Corbin, Alain. [1994] 1998. *Village Bells: Sound and Meaning in the 19th Century French Countryside*. Trans. Martin Thom. New York: Columbia University Press.

Crary, Jonathan. 2013. *24/7: Late Capitalism and the Ends of Sleep*. London: Verso.

Cubitt, Sean. 2017. *Finite Media: Environmental Implications of Digital Technologies*. Durham, NC: Duke University Press.

Cusick, Suzanne G. 2008. "You Are in a Place that is Out of the World: Music in the Detention Camps of the Global War on Terror." *Journal for the Society of American Music* 2 (1): 1–26.

Davidson, Michael. 2021. "Siting Sound: Redistributing the Senses in Christine Sun Kim." *Journal of Literary and Cultural Disability Studies* 15 (2): 219–37.

Davis, Lennard J., ed. 2013. *The Disability Studies Reader*. 4th ed. New York: Routledge.

Dery, Mark, ed. 1994. *Flame Wars: The Discourse of Cyberculture*. Durham, NC: Duke University Press.

Dittrich, Joshua. 2017. "Toward an Aural Aesthetics of 24/7 Environments: Beethoven, Audio Stretching, and Techno-Indeterminacy." *Journal of Aesthetics & Culture*. 9 (1).

Dittrich, Joshua. 2022. "Listening at Night: Toward an Ethnography of So(m)niferous Media." *Ethnologies* 44 (1): 255–72.

Dittrich, Joshua. 2023a. "Bedtime Stories: Podcasts, Audiobooks, and Reading as Listening (and Sleeping)," *Intermedialities / Intermédialités* 41: 1–20.

Dittrich, Joshua. 2023b. "The Etymology of the World: Sun Ra, Geology, Poetry." *SubStance* 52 (3): 79–94.

Dombois, Florian. 2011. "Auditory Seismology." Last modified March 15, 2011. Available online: http://www.auditory-seismology.org/version2004/ (accessed February 12, 2024).

Doughty, Karolina, Michelle Duffy, and Theresa Harada, eds. 2019. *Sounding Places: More-Than-Representational Geographies of Sound*. Cheltenham: Edward Elgar Publishing.

Droumeva, Milena and Randolf Jordan, eds. 2019. *Sound, Media, Ecology*. Cham: Palgrave MacMillan.
Dyson, Frances. 2009. *Sounding New Media: Immersion and Embodiment in the Arts and Culture*. Berkeley, CA: University of California Press.
Dyson, Frances. 2014. *The Tone Of Our Times: Sound, Sense, Economy, and Ecology*. Cambridge, MA: MIT Press.
Edwards, Brent Hayes. 2017. *Epistrophies: Jazz and the Literary Imagination*. Cambridge, MA: Harvard University Press.
Ehrenberg, Alain. 2010. *The Weariness of the Self: Diagnosing the History of Depression in the Contemporary Age*. Montreal: McGill-Queen's University Press.
Ellcessor, Elisabeth and Bill Kirkpatrick, eds. 2017. *Disability Media Studies*. New York: New York University Press.
Eno, Brian. [1978] 2017. "Ambient Music." In Christoph Cox and Daniel Warner (eds.), *Audio Culture: Readings in Modern Music*, rev. ed., 109–12. London: Bloomsbury.
Eshun, Kodwo. 1998. *More Brilliant Than the Sun: Adventures in Sonic Fiction*. London: Quartet Books.
Evens, Aden. 2005. *Sound Ideas: Music, Machines and Experience*. Minneapolis, MN: University of Minnesota Press.
Featherstone, Mike ed. 2000. *Body Modification*. London: Sage Publications.
Feld, Steven. 2015. "Acoustemology." In David Novak and Matt Sakakeeny (eds.), *Keywords in Sound*, 12–21. Durham, NC: Duke University Press.
Foster, Hal. 2004. *Prosthetic Gods*. Cambridge, MA: MIT Press.
Frantti, G. E. and L. A. Leverault. 1965. "Auditory Discrimination of Seismic Signals from Earthquakes and Explosions." *Bulletin of the Seismological Society of America* 55: 1–25.
Friedner, Michelle and Stefan Helmreich. 2012. "Sounds Studies Meet Deaf Studies." *Senses & Society* 7 (1): 72–86.
Gallagher, Michael. 2016. "Sound as Affect: Difference, Power and Spatiality." *Emotion, Space & Society* 20: 42–8.
Galloway, Alexander. 2012. *The Interface Effect*. Cambridge: Polity Press.
Garland-Thomson, Rosemarie. 2017. *Extraordinary Bodies: Figuring Physical Disability in American Culture and Literature*. 20th anniversary ed. New York: Columbia University Press.
Gilmurray, Jonathan. 2017. "Ecological Sound Art: Steps towards a New Field." *Organized Sound* 22 (1): 32–41.
Glennie, Evelyn. [2015] 2017. "Hearing Essay." In Christoph Cox and Daniel Warner (eds.), *Auditory Cultures: Readings in Modern Music*, 125–7. New York: Bloomsbury.
Glissant, Édouard. [1990] 1997. *The Poetics of Relation*. Trans. Betsy Wing. Ann Arbor, MI: University of Michigan Press.
Goodman, Steve. 2012. *Sonic Warfare: Sound, Affect and the Ecology of Fear*. Cambridge, MA: MIT Press.

Gould, Stephen Jay. 1987. *Time's Arrow, Time's Cycle: Myth and Metaphor in the Discovery of Geological Time*. Cambridge, MA: Harvard University Press.

Grosz, Elisabeth. 2008. *Chaos, Territory, Art: Deleuze and the Framing of the Earth*. New York: Columbia University Press.

Hagood, Mack. 2017. "Disability and Biomediation: Tinnitus as Phantom Disability." In Elisabeth Ellcessor and Bill Kirkpatrick (eds.), *Disability Media Studies*, 311–29. New York: New York University Press.

Hagood, Mack. 2019. *Hush: Media and Sonic Self-Control*. Durham, NC: Duke University Press.

Hansen, Mark. 2004. *New Philosophy for New Media*. Cambridge, MA: MIT Press.

Hansen, Mark. 2006. *Bodies in Code: Interfaces with Digital Media*. New York: Routledge.

Haraway, Donna. 1990. "A Cyborg Manifesto: Science, Technology and Socialist-Feminism in the Late Twentieth Century." In *Simians, Cyborgs and Women: The Reinvention of Nature*, 149–82. New York: Routledge.

Harbisson, Neil. 2010. [Video] "Sound Portraits." YouTube, 0.52. Posted by Neil Harbisson, May 7, 2010. Available online: https://www.youtube.com/watch?v=JDqL-PUZ148 (accessed June 29, 2018).

Harbisson, Neil. 2011. [Video] "Harbisson's Sonochromatic Music Scale." YouTube, 2:24. Posted by Neil Harbisson, July 20, 2011. Available online: https://www.youtube.com/watch?v=UawXwCpwDjo (accessed June 29, 2018).

Harbisson, Neil. 2012. "I Listen to Colour." TED talk, 9:29. Filmed June 2012. Available online: https://www.ted.com/talks/neil_harbisson_i_listen_to_color?language=en (accessed June 29, 2018).

Harbisson, Neil. 2013. [Video] "Colour Concert: Neil Harbisson at TEDMEDLive Imperial College." YouTube, 8:29. Posted by TEDMEDLive, May 23, 2013. Available online: https://www.youtube.com/watch?v=6P8O5JXlAJg (accessed June 29, 2018).

Harbisson, Neil. 2015a. "Den Farben der Sonne lauschen." Interview by Silvia Cachafeiro and Martin Zingl. Posted by Wiener Zeitung, July 24, 2015. Available online: https://www.wienerzeitung.at/themen_channel/wz_reflexionen/zeitgenossen/?em_cnt=765064 (accessed June 29, 2018).

Harbisson, Neil. 2015b. [Video] "Neil Harbisson First Colour-Conducted Concert." YouTube, 10:51. Posted by Me and You Films, October 23, 2015. Vodaphone Sponsored Content. Available online: https://www.youtube.com/watch?v=lsj_LADL_Gg (accessed June 29, 2018).

Harbisson, Neil. 2017. [Video] "Neil Harbisson on Being a Cyborg." YouTube, 8:41. Posted by HSGUniStGallen, May 9, 2017. Available online: https://www.youtube.com/watch?v=C_OnYqx3ynA (accessed June 29, 2018).

Harbisson, Neil. 2018. "Sonochromatism." Wikipedia. Available online: https://en.wikipedia.org/wiki/Sonochromatism (accessed June 29, 2018).

Harbisson, Neil. Cyborg Foundation, website. Available online: https://www.cyborgfoundation.com (accessed November 1, 2017).

Harbisson, Neil. Transpecies Society, website. Available online: https://www.transpeciessociety.com (accessed June 29, 2018).

Harbisson, Neil and Moon Ribas. 2016. "What's It Like to Be a Cyborg?" YouTube, 1.00:47. Posted by Talks at Google, November 21, 2016. Available online: https://www.youtube.com/watch?v=rRU62Csr_jI (accessed June 29, 2018).

Harman, Graham. 2002. *Tool-Being: Heidegger and the Metaphysics of Objects*. Chicago, IL: Open Court.

Havis, Devonya N. 2023. *Creating a Black Vernacular Philosophy*. Lanham, MD: Lexington Books.

Hayles, N. Katherine. 1999. *How We Became Posthuman: Virtual Bodies in Cybernetics, Literature and Informatics*. Chicago, IL: University of Chicago Press.

Hayles, N. Katherine. 2017. *Unthought: The Power of the Cognitive Nonconscious*. Chicago, IL: University of Chicago Press.

Hayward, Chris. 1994. "Listening to the Earth Sing." In Gregory Kramer (ed.), *Auditory Display: Sonification, Audification, and Auditory Interfaces*, Proceedings Volume XVIII Santa Fe Institute, 369–404. Reading, MA: Addison-Wesley.

Heidegger, Martin. [1936] 1977. "The Origin of the Work of Art." In David Farrell Krell (ed.), *Basic Writings*, 149–87. New York: Basic Books.

Heller-Roazen, Daniel. 2011. *The Fifth Hammer: Pythagoras and the Disharmony of the World*. New York: Zone Books.

Helmreich, Stefan. 2007. "An Anthropologist Underwater: Immersive Soundscapes, Submarine Cyborgs, and Transductive Ethnography." *American Ethnologist* 34 (4): 621–41.

Helmreich, Stefan. 2015. "Transduction." In David Novak and Matt Sakakeeny (eds.), *Keywords in Sound*, 222–31. Durham, NC: Duke University Press.

Hennesey, Kate. 2016. "Max Richter's *Sleep* Review: Exquisitely Soundtracked Sleepover at the Sydney Opera House." *The Guardian*, June 6. Available online: https://www.theguardian.com/culture/2016/jun/06/max-richters-sleep-review-an-exquisitely-soundtracked-sleepover-at-the-sydney-opera-house (accessed July 15, 2023).

Hersey, Tricia. 2022. *Rest is Resistance: A Manifesto*. Little Brown, Spark.

Hill, Edwin C. Jr. 2013. *Black Soundscapes, White Stages: The Meaning of Francophone Sound in the Black Atlantic*. Baltimore, D: Johns Hopkins University Press.

Horton, Zachary. 2021. *The Cosmic Zoom: Scale, Knowledge, and Mediation*. Chicago, IL: University of Chicago Press.

Hosokawa, Shuhei. 2012. "The Walkman Effect." In Jonathan Sterne (ed.), *The Sound Studies Reader*, 104–16. New York: Routledge.

Inge, Leif. 2002. [Sound installation] *9 Beet Stretch*.

Ingold, Tim. 2007. *Autumn Leaves: Sound and the Environment in Artistic Practice*. Ed. by Angus Carlyle, 10–13. Paris: Double Entendre.

Jackson, Chandra L., Jenelle R. Walker, Marishka K. Brown, and Nancy Lynne Jones. 2020. "A Workshop Report on the Causes and Consequences of Sleep Health Disparities." *Sleep* 43 (8).

Jackson, Chandra L., Sanjay R. Patel, W. Braxton Jackson, Pamela L. Lutsey, and Susan Redline. 2018. "Agreement between Self-Reported and Objectively Measured Sleep Duration among White, Black, Hispanic, and Chinese Adults in the US: Multi-Ethnic Study of Atherosclerosis." *Sleep* 41 (6).

James, Robin. 2019. *The Sonic Episteme: Acoustic Resonance, Neoliberalism, and Biopolitics*. Durham, NC: Duke University Press.

Jue, Melody. 2020. *Wild Blue Media: Thinking Through Seawater*. Durham, NC: Duke University Press.

Kahn, Douglas. 1997. "John Cage: Silence and Silencing." *Musical Quarterly* 81 (4): 556–98.

Kahn, Douglas. 1999. *Noise, Water, Meat: A History of Sound in the Arts*. Cambridge, MA: MIT Press.

Kahn, Douglas. 2004. "Ether Ore: Mining Vibrations in American Modernist Music." In Veit Erlmann (ed.), *Hearing Cultures: Essays on Sound, Listening and Modernity*, 107–30. New York: Berg.

Kahn, Douglas. 2011. "The Latest: Fluxus and Music." In Caleb Kelly (ed.), *Sound: Documents of Contemporary Art*, 28–42. London: Whitechapel Gallery.

Kahn, Douglas. 2013. *Earth Sound, Earth Signal: Energies and Earth Magnitude in the Arts*. Berkeley, CA: University of California Press.

Kassabian, Annahid. 2013a. "Music for Sleeping." In Marie Thompson and Ian Biddle (eds.), *Sound, Music, Affect: Theorizing Sonic Experience*, 165–81. London: Bloomsbury.

Kassabian, Annahid. 2013b. *Ubiquitous Listening: Affect, Attention and Distributed Subjectivity*. Berkeley, CA: University of California Press.

Keeling, Kara. 2019. *Queer Times, Black Futures*. New York: New York University Press.

Kelly, Caleb, ed. 2011. *Sound: Documents of Contemporary Art*. London: Whitechapel.

Kim, Christine Sun. 2013. [Video] "Courtier as Courier: Voiceless Lecture," 3:14, posted on Vimeo by the artist. Available online: https://vimeo.com/68031095 (accessed December 12, 2023).

Kim, Christine Sun. 2015. "The Enchanting Music of Sign Language." TED talk. Available online: https://www.ted.com/talks/christine_sun_kim_the_enchanting_music_of_sign_language?language=en (accessed July 18, 2023).

Kim-Cohen, Seth. 2009a. *In the Blink of an Ear: Toward a Non-Cochlear Sonic Art*. New York: Continuum.

Kim-Cohen, Seth. 2009b. "The Hole Truth: Seth Kim-Cohen on Doug Aitken's *Sonic Pavilion*." *Artforum International*, November 2009, 99+. *Gale in Context: Biography*. Available online: https://link-gale-com.myaccess.library.utoronto.ca/apps/doc/A211807973/BIC?u=utoronto_main&sid=BIC&xid=81fd0b00 (accessed March 12, 2020).

Kittler, Friedrich. [1986] 1999. *Gramophone, Film, Typewriter*. Trans. with an introduction, by Geoffrey Winthrop-Young and Michael Wutz. Stanford, CA: Stanford University Press.

Kittler, Friedrich. 2006. "Lightning and Series—Event and Thunder." Trans. Geoffrey Winthrop-Young. *Theory, Culture & Society* 23 (7–8): 63–74.

Knoll, Andrew. 2021. *A Brief History of Earth: Four Billion Years in Eight Chapters*. New York: Mariner Books.

Kramer, Gregory. 1994. "An Introduction to Auditory Display." In Gregory Kramer (ed.), *Auditory Display: Sonification, Audification, and Auditory Interfaces*, Proceedings Volume XVIII Santa Fe Institute, 1–79. Reading, MA: Addison-Wesley.

Kroker, Kenton. 2007. *The Sleep of Others and the Transformations of Sleep Research*. Toronto: University of Toronto Press.

Kroker, Arthur and Marilouise Kroker, eds., 2008. *Critical Digital Studies: A Reader*, 442–512. Toronto: University of Toronto Press.

LaBelle, Brandon. [2010] 2019. *Acoustic Territories: Sound Culture and Everyday Life*, 2nd ed. New York: Bloomsbury.

Lanza, Joseph. 2004. *Elevator Music: A Surreal History of Muzak®, Easy-Listening, and Other Moodsong®*, rev. and expanded ed. Ann Arbor, MI: University of Michigan Press.

Lock, Graham. 1999. *Blutopia: Visions of the Future and Revisions of the Past in the Work of Sun Ra, Duke Ellington, and Anthony Braxton*. Durham, NC: Duke University Press.

Lockley, Steven W. and Russell G. Foster. 2012. *Sleep: A Very Short Introduction*. Oxford: Oxford University Press.

Mackenzie, Adrian. 2002. *Transductions: Bodies and Machines at Speed*. New York: Continuum.

Malabou, Catherine. 2008. *What Should We Do with Our Brain?* Foreword by Marc Jeannerod, trans. Sebastian Rand. New York: Fordham University Press.

Martin, Allie. Black Sound Lab. www.blacksoundlab.com. Digital Humanities and Social Engagement Cluster. Dartmouth College, Hanover, NH.

Massumi, Brian. 2002. *Parables for the Virtual: Movement, Affect, Sensation*. Durham, NC: Duke University Press.

Maxwell, Richard and Toby Miller. 2012. *Greening the Media*. New York: Oxford University Press.

McCormack, Derek. 2018. *Atmospheric Things: On the Allure of Elemental Envelopment*. Durham, NC: Duke University Press.

McGee, Ryan. 2012. "Haiti Earthquake" and "Christchurch Earthquake." "Seismic Sounds: Seismic Auditory Display." Available online: https://soundcloud.com/seismicsounds (accessed March 27, 2020).

McGee, Ryan and David Rogers. 2013. "Sounds of Seismic: Earth System Soundscape." Available online: http://sos.allshookup.org (accessed February 28, 2020).

McGee, Ryan and David Rogers. 2016. "Musification of Seismic Data." Paper presented at the 22nd International Conference on Auditory Display, Canberra, July 2–8.

McPhee, John. 1998. *Annals of the Former World*. New York: Farrar, Straus and Giroux.

McRuer, Robert. 2006. *Crip Theory: Cultural Signs of Queerness and Disability*. Foreword by Michael Bérubé. New York: New York University Press.

Miles, Kathryn. 2017. *Quakelands: On the Road to America's Next Devastating Earthquake*. New York: Dutton.

Mills, Mara. 2015. "Deafness." In David Novak and Matt Sakakeeny (eds.), *Keywords in Sound*, 45–54. Durham, NC: Duke University Press.

Mitchell, David T. and Sharon L. Snyder. 2001. *Narrative Prosthesis: Disability and the Dependencies of Discourse*. Ann Arbor, MI: University of Michigan Press.

Mitchell, David T., Susan Antebi, and Sharon L. Snyder, eds. 2019. *The Matter of Disability: Materiality, Biopolitics, Crip Affect*. Ann Arbor, MI: University of Michigan Press.

Mitchell, W. J. T. 2005. "There Are No Visual Media." *Journal of Visual Culture* 4 (2): 257–66.

Monahan, Gordon. 2010. "Canada Council Laureate 2013 Gordon Monahan, A Piano Listening To Itself – Chopin Chord, 2010." Available online: https://www.youtube.com/watch?v=YUi8MNRhXT8 (accessed February 28, 2020).

More, Max and Natasha Vita-More, eds. 2013. *The Transhumanist Reader: Classical and Contemporary Essays on the Science, Technology, and Philosophy of the Human Future*. Malden, MA: Wiley-Blackwell.

Morton, Timothy. 2013. *Hyperobjects: Philosophy and Ecology after the End of the World*. Minneapolis, MN: University of Minnesota Press.

Mumma, Gordon. 2007. [CD] *David Tudor and Gordon Mumma*. New World Records.

Mumma, Gordon. 2008. [CD] *Music for Solo Piano 1960–2001*. Performed by Daan Vandewalle. New World Records.

Nancy, Jean-Luc. 2009. *The Fall of Sleep*. Trans. Charlotte Mandell. New York: Fordham University Press.

Nelson, Alondra. 2002. "Introduction: Future Texts." *Social Text* 20 (2): 1–15.

Novak, David and Matt Sakakeeny. 2015. "Introduction." In David Novak and Matt Sakakeeny (eds.), *Keywords in Sound*, 1–11. Durham, NC: Duke University Press.

O'Connell, Mark. 2017. *To Be a Machine: Adventures Among Cyborgs, Utopians, Hackers and the Futurists Solving the Modest Problem of Death*. New York: Anchor Books.

Oliveros, Pauline. [2011] 2017. "Auralizing the Sonosphere: A Vocabulary for Inner Sound and Sounding." In Christoph Cox and Daniel Warner (eds.), *Audio Culture: Readings in Modern Music*, 2nd ed., 113–16. New York: Bloomsbury.

Omelsky, Matthew. 2023. *Fugitive Time: Global Aesthetics and the Black Beyond*. Durham, NC: Duke University Press.

Oster, Gerald. 1973. "Auditory Beats in the Brain." *Scientific American* 224 (4): 94–103.

Ouzounian, Gascia and Keith and Mendi Obadike. 2022. Sleep Salon 7: Sound and Sleep, Montreal-based research project Sociability of Sleep, May 6. Available online: https://www.youtube.com/watch?v=fRCNZ6jysnI (accessed July 15, 2023)

Paglen, Trevor. 2012a. "The Last Pictures: Interview with Trevor Paglen." Interviewed by Nato Thompson. *e-flux* 37.

Paglen, Trevor. 2012b. *The Last Pictures*. New York and Berkeley, CA: Creative Time Books and University of California Press.

Parikka, Jussi. 2015. *A Geology of Media*. Minneapolis, MN: University of Minnesota Press.

Parry-Davies, Ella. 2016. "*24/7: Late Capitalism and the Ends of Sleep* by Jonathan Crary (Review)." *Cinema Journal* 55 (2): 177–81.

Paterson, Katie. 2021. "The Mind-Bending Art of Deep Time." TED talk. Available online: https://www.ted.com/talks/katie_paterson_the_mind_bending_art_of_deep_time (accessed July 10, 2023).

Pesic, Peter. 2014. *Music and the Making of Modern Science*. Cambridge, MA: MIT Press.

Peters, John Durham. 2015. *The Marvelous Clouds: Toward a Philosophy of Elemental Media*. Chicago, IL: University of Chicago Press.

Peterson, Marina. 2021. *Atmospheric Noise: The Indefinite Urbanism of Los Angeles*. Durham, NC: Duke University Press.

Pynchon, Thomas. [1973] 2000. *Gravity's Rainbow*. New York: Penguin Books.

Rabinbach, Anson. 1990. *The Human Motor: Energy, Fatigue and the Origins of Modernity*. Berkeley, CA: University of California Press.

Raffles, Hugh. 2020. *The Book of Unconformities: Speculations on Lost Time*. New York: Pantheon Books.

Reiss, Benjamin. 2017. *How Taming Sleep Created Our Restless World*. New York: Basic Books.

Revill, Georg. 2016. "How is Space Made in Sound? Spatial Mediation, Critical Phenomenology and the Political Agency of Sound." *Progress in Human Geography* 40 (2): 240–56.

Ribas, Moon. 2013. "Waiting for Earthquakes." Uploaded by Moon Ribas, April 4, 2013. Available online: https://www.youtube.com/watch?v=1Un4MFR-vNI (accessed February 28, 2020).

Ribas, Moon. 2015. "Searching for My Sense." TEDxMünchen. Uploaded January 27, 2015. Available online: https://www.youtube.com/watch?v=qU6UPUlbmLw (accessed February 28, 2020).

Ribas, Moon. 2016. "Earthbeat." TEDxMcGill. Uploaded April 14, 2016. Available online: https://www.youtube.com/watch?v=oPuJ4ucXxAQ (accessed February 28, 2020).

Ribas, Moon. 2017. "Moon Ribas: The Dancer Whose Arm Vibrates Every Time There's an Earthquake." Available online: https://www.youtube.com/watch?v=A8o9ISOgLBc (accessed February 20, 2020).

Richter, Max. 2015a. [CD] From *Sleep*, with Grace Davidson (soprano) and American Contemporary Music Ensemble. Deutsche Grammophon, 2015.

Richter, Max. 2015b. [CD] *Sleep*, with Grace Davidson (soprano) and American Contemporary Music Ensemble. Deutsche Grammophon, 2015.

Richter, Max. 2018. "Max Richter's *Sleep* – Sydney Opera House." Posted by MaxRichterMusic, July 22, 2018. Available online: https://www.youtube.com/watch?v=lHMCE-c8sUc (accessed December 2, 2019).

Rider, Alexis and Paul A. Harris. 2023. "Introduction: Breaking Earth." *SubStance* 52 (3): 3–8.

Robinson, Dylan. 2020. *Hungry Listening: Resonant Theory for Indigenous Sound Studies.* Minneapolis, MN: University of Minnesota Press.

Roden, Steve. 2004. *ear(th).* Catalogue (with audio CD) to the sound installation. Pasadena, CA: Alyce de Roulet Williamson Gallery / Art Center College of Design.

Ronell, Avital. 1989. *The Telephone Book: Technology, Schizophrenia, Electric Speech.* Lincoln, NE: University of Nebraska Press.

Rose, Nikolas. 2016. "Reading the Human Brain: How the Mind Became Legible." *Body & Society* 22 (2): 140–77.

Rothblatt, Martine. 2013. "Mind is Deeper Than Matter: Transgenderism, Transhumanism, and the Freedom of Form." In Max More and Natasha Vita-More (eds.), *The Transhumanist Reader: Classical and Contemporary Essays on the Science, Technology, and Philosophy of the Human Future*, 317–26. Malden, MA: Wiley-Blackwell.

Sandberg, Anders. 2013. "Morphological Freedom – Why We Not Just Want It, but Need It." In Max More and Natasha Vita-More (eds.), *The Transhumanist Reader: Classical and Contemporary Essays on the Science, Technology, and Philosophy of the Human Future*, 56–64. Malden, MA: Wiley-Blackwell.

Schafer, R. Murray. 1977. *The Soundscape: Our Sonic Environment and the Tuning of the World.* Rochester, VT: Destiny Books.

Schwarz, Hillel. 2011. *Making Noise: From Babel to the Big Bang & Beyond.* New York: Zone Books.

Sharma, Sarah. 2014. *In the Meantime: Temporality and Cultural Politics.* Durham, NC: Duke University Press.

Siebers, Tobin. 2008. *Disability Theory.* Ann Arbor, MI: University of Michigan Press.

Siebers, Tobin. 2010. *Disability Aesthetics.* Ann Arbor, MI: University of Michigan Press.

Simmel, Georg. 1997. "Metropolis and Mental Life." In Hans Gerth (trans.) and David Frisby and Mike Featherstone (eds.), *Simmel on Culture*, 174–85. London: Sage Publications.

Simondon, Gilbert. 1992. "The Genesis of the Individual." In Mark Cohen and Sanford Kwinter (trans.) and Jonathan Crary and Sanford Kwinter (eds.), *Incorporations*, 296–319. New York: Zone Books.

Sites, William. 2020. *Sun Ra's Chicago: Afrofuturism and the City.* Chicago, IL: University of Chicago Press.

Smith, Jacob. 2016. *Eco-Sonic Media.* Berkeley, CA: University of California Press.

Speeth, Sheridan Dauster. 1961. "Seismometer Sounds." *Journal of the Acoustical Society of America* 33 (7): 909–16.

Starosielski, Nicole and Janet Walker, eds. 2016. *Sustainable Media: Critical Approaches to Media and Environment.* New York: Routledge.

Sterne, Jonathan. 2003. *The Audible Past: Cultural Origins of Sound Reproduction.* Durham, NC: Duke University Press.

Sterne, Jonathan, ed. 2012. *The Sound Studies Reader.* New York: Routledge.

Sterne, Jonathan. 2013. "Soundscape, Landscape, Escape." In Karin Bijsterveld (ed.), *Soundscapes of the Urban Past*, 181–93. Bielefeld: transcript.

Sigrid Hauff, Klaus Detlef Thiel, and Brent Hayes Edwards. 2005. *Sun Ra: The Immeasurable Equation. The Collected Poetry and Prose.* Compiled and ed. by James L. Wolf and Hartmut Geerken. Wartaweil: Waitawhile.

Sun Ra. 1963. [CD] *Cosmic Tones for Mental Therapy* and *Art Forms of Dimensions Tomorrow*. Produced by Infinity and Alton Abraham.

Sun Ra. 1974. [Film] *Space is the Place*. Dir. John Coney. Written by Sun Ra and Joshua Smith. USA: Plexifilm, 85 minutes.

Sun Ra. 1976. "Sun Ra Reads Poetry 12/25/1976 WXPN Philadelphia," December 25. Available online: https://www.youtube.com/watch?v=TmdqjXTkrD8 (accessed June 20, 2023).

Sun Ra. 2005. *Sun Ra: The Immeasurable Equation. The Collected Poetry and Prose.* Compiled and ed. James L. Wolf and Hartmut Geerken. Introductions and Essays by James L. Wolf, Hartmut Geerken,

Sun Ra. 2006. *The Wisdom of Sun Ra: Sun Ra's Polemical Broadsheets and Streetcorner Leaflets.* Compiled and introduced by John Corbett. Chicago, IL: WhiteWalls.

Sun Ra. 2011. *This Planet is Doomed: The Science Fiction Poetry of Sun Ra*. Foreword by Amiri Baraka. New York: Kicks Books.

Szwed, John. [1997] 2020. *Space is the Place: The Lives and Times of Sun Ra*. Durham, NC: Duke University Press.

Tan, Eliza. 2020. "Voicing the Sociality of Sound from a Deaf Perspective: Christine Sun Kim's Plenitude of Silence." *Oxford Art Journal* 43 (2): 239–60.

Thompson, Emily. [2002] 2012. "Sound, Modernity, History." In Jonathan Sterne (ed.), *The Sound Studies Reader*. 117–29. New York: Routledge.

Thompson, Marie. 2017. *Beyond Unwanted Sound: Noise, Affect, and Aesthetic Moralism*. New York: Bloomsbury.

Thompson, Marie and Ian Biddle, eds. 2013. *Sound, Music, Affect: Theorizing Sonic Experience*. London: Bloomsbury.

Toop, David. [1995] 2018. *Ocean of Sound: Ambient Sound and Radical Listening in the Age of Communication*. London: Serpent's Tail.

Tremain, Shelly, ed. 2015. *Foucault and the Government of Disability*. Enlarged and rev. ed. Ann Arbor, MI: University of Michigan Press.

Turner, Bryan S., ed. 2012. *Routledge Handbook of Body Studies*. New York: Routledge.

Veal, Michael. 2007. *Dub: Soundscapes and Shattered Songs in Jamaican Reggae*. Middleton, CT: Wesleyan University Press.

Westerkamp, Hildegard. 2019. "The Disruptive Nature of Listening: Today, Yesterday, Tomorrow." In Milena Droumeva and Randolf Jordan (eds.), *Sound, Media, Ecology*, 45–63. Cham: Palgrave MacMillan.

Williams, Simon J. 2011. *The Politics of Sleep: Governing (Un)consciousness in the Late Modern Age*. Houndmills: Palgrave Macmillan.

Wills, David. 1995. *Prosthesis*. Stanford, CA: Stanford University Press.

Wolbring, Gregor. 2006. "The Enhancement Debate, or Ableism Leads to Transhumanism." Conference Presentation, "Tomorrow's People: The Challenges of Technologies for Life Extension and Enhancement," Said Business School, University of Oxford. Written retrieved on July 19, 2023. Available online: https://cspo.org/legacy/library/091104F4KN_lib_Wolbring2006pres.pdf (accessed July 19, 2023).

Wolf-Meyer, Matthew. 2012. *The Slumbering Masses: Sleep, Medicine, and Modern American Life*. Minneapolis, MN: University of Minnesota Press.

Woodard, Ben. 2013. *On an Ungrounded Earth: Towards a New Geophilosophy*. Brooklyn, CA: Punctum Books.

Youngquist, Paul. 2016. *A Pure Solar World: Sun Ra and the Birth of Afrofuturism*. Austin, TX: University of Texas Press.

Yusoff, Kathryn. 2015. "Geologic Subjects: Non-Human Origins, Geomorphic Aesthetics and the Art of Becoming *In*human." *Cultural Geographies* 22 (3): 383–407.

Yusoff, Kathryn. 2018. *A Billion Black Anthropocenes or None*. Minneapolis, MN: University of Minnesota Press.

Zielinski, Siegfried. [2002] 2006. *Deep Time of the Media: Toward an Archaeology of Hearing and Seeing by Technical Means*. Trans. Gloria Custance. Cambridge, MA: MIT Press.

Zylinska, Joanna, ed. 2002. *The Cyborg Experiments: The Extensions of the Body in the Media Age*. New York: Continuum.

Index

Page numbers in **bold** refer to figures.

www.ingramcontent.com/pod-product-compliance
Ingram Content Group UK Ltd.
Pitfield, Milton Keynes, MK11 3LW, UK
UKHW020125290926
15471UKWH00003B/283

9 798765 104569